HOW TO END A WAR

How and when should we end a war? What place should the pathways to a war's end have in war planning and decision-making? This volume treats the topic of ending war as part and parcel of how wars begin and how they are fought – a unique, complex problem, worthy of its own conversation. New essays by leading thinkers and practitioners in the fields of philosophical ethics, international relations, and military law reflect on the problem and show that it is imperative that we address not only the resolution of war, but how and if a war as waged can accommodate a future peace. The essays collectively solidify the topic and underline its centrality to the future of military ethics, strategy, and war.

GRAHAM PARSONS is Associate Professor of Philosophy at United State Military Academy, West Point.

MARK A. WILSON is Professor of Ethics at Villanova University.

HOW TO END A WAR

Essays on Justice, Peace, and Repair

EDITED BY

GRAHAM PARSONS

United States Military Academy, West Point

MARK A. WILSON

Villanova University

CAMBRIDGE
UNIVERSITY PRESS

CAMBRIDGE
UNIVERSITY PRESS

Shaftesbury Road, Cambridge CB2 8EA, United Kingdom

One Liberty Plaza, 20th Floor, New York, NY 10006, USA

477 Williamstown Road, Port Melbourne, VIC 3207, Australia

314–321, 3rd Floor, Plot 3, Splendor Forum, Jasola District Centre, New Delhi – 110025, India

103 Penang Road, #05–06/07, Visioncrest Commercial, Singapore 238467

Cambridge University Press is part of Cambridge University Press & Assessment, a department of the University of Cambridge.

We share the University's mission to contribute to society through the pursuit of education, learning and research at the highest international levels of excellence.

www.cambridge.org
Information on this title: www.cambridge.org/9781108834285

DOI: 10.1017/9781108992640

First published 2023

A catalogue record for this publication is available from the British Library.

ISBN 978-1-108-83428-5 Hardback

Contents

Contributors

ALEX J. BELLAMY is Director of the Asia Pacific Centre for the Responsibility to Protect and Professor of Peace and Conflict Studies at the University of Queensland, Australia.

NIR EISIKOVITS is Associate Professor of Philosophy and founding Director of the Applied Ethics Center at the University of Massachusetts, Boston.

DAN MAURER is a lieutenant colonel in the U.S. Army and currently Assistant Professor of Law at the United States Military Academy, West Point.

DARREL MOELLENDORF is Professor of International Political Theory and Professor of Philosophy at Johann Wolfgang Universität, Frankfurt am Main.

COLLEEN MURPHY is the Roger and Stephany Joslin Professor of Law and Professor of Philosophy and Political Science at the University of Illinois at Urbana-Champaign.

GRAHAM PARSONS is Associate Professor in the Department of English and Philosophy at the United States Military Academy, West Point.

DANIEL PHILPOTT is Professor of Political Science at the University of Notre Dame.

DAVID RODIN is a senior research fellow at the University of Oxford and Senior Fellow at the Carnegie Council for Ethics in International Affairs.

CHEYNEY RYAN is Senior Research Fellow at the Institute for Ethics, Law, and Armed Conflict at the University of Oxford.

NANCY SHERMAN is University Professor of Philosophy at Georgetown University.

LISA TESSMAN is Professor of Philosophy at Binghamton University.

MARK A. WILSON is a Teaching Professor of Ethics in the Ethics Program at Villanova University.

Acknowledgments

This volume grew out of a conference that took place at West Point in the fall of 2019. The theme was inspired by the 100th anniversary of the signing of the Treaty of Versailles and its notorious example of how not to end a war. This book would not have been possible without those who made that conference possible. We would like to thank Tony Baker, Thomas Carnes, Reece Doty, Hunter Cantrell, Kris Fox, Terron Johnson, Caroline Pagan, Gary Bridges, and Janice Torreta. West Point's Department of English and Philosophy funded the conference. The Association of Graduates and the Center for Holocaust and Genocide Studies at West Point provided logistical support. Thanks also to Mark Doorley, Director of Villanova's Ethics Program, who has shown unwavering enthusiasm and backing for these collaborations.

Hilary Gaskin at Cambridge University Press has been patient and supportive of this project from the beginning. We are grateful for her vision and guidance.

We are most appreciative of the outstanding contributors to this volume. It is rare that one finds a group of scholars so collegial, open-minded, and mutually supportive. It has been inspiring to work with them. We wish everyone in the field were half as generous and kind. They deserve all credit for this book's quality.

Introduction
The Ethics of War after the Longest War

Graham Parsons and Mark A. Wilson

As we write, it has been four months since the United States formally ended its military operations in Afghanistan. The war was a disaster in many respects. Western military planners badly misjudged their ability to achieve their objectives and the war dragged on far longer than predicted, lasting almost exactly 20 years and becoming the longest war in American history. One of the initial goals of the war was to eliminate the Taliban as a political force in Afghanistan. However, in scenes reminiscent of the end of its war in Vietnam, U.S. forces ended their war with a chaotic evacuation, as the group it had aimed to destroy retook control of most of the country. Of course, what makes failed wars like this most terrible is their human toll. In addition to failing to achieve its most lofty goals, the war in Afghanistan destroyed countless lives along the way.

Sadly, America's war in Afghanistan is emblematic of the character of all its major military engagements since September 11, 2001. In Iraq too, the U.S. led an invasion and occupation that lasted far longer than anticipated and produced overwhelming tragedy for Iraq and its neighboring countries. And in Libya, the U.S. joined NATO forces in an invasion to prevent atrocities and topple a dictatorship, but unintentionally produced a protracted civil war with vast casualties.

Additionally, recall that these wars (with the possible exception of Libya) are part of a larger war – the global war on terror. This war endures despite the end of the war in Afghanistan and will do so even if closure is ultimately found in Iraq and Libya. The threat of international terrorism by nonstate actors continues and their number has arguably increased since September 11. The war on terror is yet another endless American war.

Certainly, these catastrophes reveal the shortcomings of the political and military leaders who conceived of and executed these wars, and who are the agents of the multitude of errors that compose them. Still, those of us who work on the ethics of war face a reckoning of our own. We must admit that the toolkit that we provide to political and military leaders in the form of conventional military ethics and law has proved inadequate for the current crises and has perhaps even contributed to these tragedies.

Consider the U.S.-led war in Afghanistan. The basic components of just war theory did not provide the guidance leaders needed to make better decisions. The war did not clearly violate the conditions of *jus ad bellum*. Unlike the debates that swirled around the decision to invade Iraq, there was a reasonable argument that the invasion of Afghanistan had a just cause. Indeed, one of the leading textbooks of military ethics treats the invasion of Afghanistan as an example of justified resistance to aggression.[1] The constraints imposed by *jus in bello* offered limits on the tactics leaders could employ in Afghanistan. These limits were sometimes transgressed and *jus in bello* was a useful resource for criticism in these cases. But an unexpected thing happened in Afghanistan (and Iraq): As the strategic challenge of winning a counterinsurgency became clearer, the conventions of *jus in bello* proved inadequate and rules of engagement were adopted that were much more restrictive. For instance, in many respects, American warfighters eventually took greater effort to avoid noncombatant casualties than conventional ethics and law require.[2] *Jus in bello*, while not rejected, was made redundant by the strategic necessities of this war.

This simultaneous dovetailing and divergence of military strategy and *jus in bello* brings to light a serious gap in just war thinking. Just as the experience of war in Afghanistan showed military planners that there is more to ending a war than overwhelming an opponent militarily, that same lesson needs to be embraced by just war theorists. The basic structure of just war theory is built around the assumption – the same assumption that gripped those who instigated the war in Afghanistan – that ending a war is not a strategically complex act. In the simplest terms, the assumption is that once an enemy force is sufficiently crushed, its leaders will accept political demands and peace will be achieved. The standards of *jus ad bellum* and *jus in bello* restrict mainly what political demands you can make in war and the methods used to defeat the enemy in the pursuit of those demands. The theory falls silent when there is no longer a strategic connection between increasing attacks on opposing forces and victory. For the extraordinary ethical challenges that one faces when victory is not a matter of mere destruction but of political negotiation and compromise under pressure of military action, of weaving military action together with

[1] Brian Orend, *The Morality of War*, second edition (Buffalo, NY: Broadview Press, 2013).

[2] Joseph Felter and Jacob Shapiro, "Limiting Civilian Casualties as Part of a Winning Strategy: The Case of Courageous Restraint," *Daedalus*, 146, no. 1 (2017), 44–58.

diplomacy, development, and economics – the very challenges that the American military faced in Afghanistan for so long – just war theory has been little help.

This problem is not solved by the inclusion of the doctrines of *jus post bellum* into just war theory. While in the last few decades there has developed a robust and influential literature in this area, *jus post bellum* presumes the end of war; it is precisely an ethic for *after* war. But the challenge in Afghanistan was how to achieve the end, how to get to a postwar situation.

Another respect in which the war in Afghanistan (and, in this case, Iraq as well) has challenged conventional military ethics and law is that it has reiterated how limited the language of just war theory is. As reflected in the mental health crisis among veterans, the toll of these disasters stretches beyond the killed and maimed, the destroyed property and livelihoods, the social and political instability. One recent study concludes that 30,177 American veterans and active-duty service members have died by suicide since 9/11, whereas 7,057 have died in military operations during the same period.[3] War wounds in many ways and surviving it requires more than avoiding enemy fire. While just war theory reduces the ethics of acts in war to the simple categories of "permissible," "impermissible," and "excused," the suffering characteristic of the experience of war suggests that a richer language is needed to capture the ethical realities. Too many service members fail to find solace in the "permissibility" of their actions.

In partial recognition of these problems, the scholarship on the ethics and laws of war has evolved in significant, but as yet underdeveloped ways. First, some scholars have tried to carve out new ethical and legal frameworks that bridge war and postwar, or what has been understood conventionally as *jus in bello* and *jus post bellum*. Darrel Moellendorf and David Rodin, two of the contributors to this volume, have coined the phrases *jus ex bello* and *jus terminatio*, respectively, to describe this framework. What they illuminate is that the issues of when and how to end a war are much more imbricated with how and why wars are fought – the *jus in bello* and *jus ad bellum* – than has been traditionally recognized. Second, there is now significant scholarship on the relationship between injustice manifest at *ad bellum* and *in bello* levels and the creation of longer-term threats to national security *post bellum*, especially threats of terrorism. Historical and structural injustices are increasingly seen as risks to national security

[3] Jennifer Steinhauer, "Suicides Among Post-9/11 Veterans Are Four Times as High as Combat Deaths, a New Study Finds," *The New York Times*, June 22, 2021, www.nytimes.com/2021/06/22/us/911-suicide-rate-veterans.html.

and, therefore, there are strategic as well as ethical reasons to address such injustices. Third, there is now widespread understanding that war, even war that follows the laws of armed conflict, poses a serious psychic danger to combatants that can cause them harm long after the shooting stops. The surging interest in the topic of moral injury is testament to this, and understanding this phenomenon requires deeper attention to the reasons for which wars are waged and how they are pursued. As the work of Jonathan Shay and Brett Litz et al. indicate, the experience of combatants "*post bellum*" is often deeply tied to concerns about the reasons for and methods of war.[4]

This volume bridges these usually disconnected conversations about when to go to war, how to wage war, and how to navigate the aftermath of war in ways that no current scholarship fully addresses. Unlike leading works in *jus post bellum*, this book treats the topic of ending war as part and parcel of how wars begin and how they are fought – a unique, complex problem, worthy of its own conversation. When confronting the failures of America's war in Afghanistan and the prospect of wars without end, it is imperative not only that we address the resolution of war, but how and if a war as waged can accommodate a future peace. In this interdisciplinary volume, we bring together ten original chapters by some of the foremost thinkers and practitioners in the fields of just war theory, international relations, and military law. Not only do these essays individually advance the conversations in this area, but collectively they solidify the topic as central to the future of military ethics, law, and strategy.

*****

The late E.O. Wilson famously said, "The real problem with humanity is the following: we have Paleolithic emotions, medieval institutions, and god-like technology."[5] Though his target was not war, and we might question whether emotions should be submerged in the stone ages, he captures a critical disconnect between the emotions experienced by those engaged in war, the medieval origins of just war thinking, and the profound complexities borne of the contemporary machinery – and massive destructive powers – of modern warfare.

[4] See Jonathan Shay, *Achilles in Vietnam: Combat Trauma and the Undoing of Character* (New York: Simon and Schuster, 2010); and Brett T. Litz et al., "Moral Injury and Moral Repair in War Veterans: A Preliminary Model and Intervention Strategy," *Clinical Psychology Review*, 29, no. 8 (2009), 695–706.

[5] E. O. Wilson, "Debate at the Harvard Museum of Natural History" (Cambridge, MA, September 9, 2009), cited by Tristan Harris in "Our Brains Are No Match for Our Technology," *The New York Times*, December 5, 2019, www.nytimes.com/2019/12/05/opinion/digital-technology-brain.html.

The first four chapters of this volume explore the many ways that both the precipitants for and aftermath of war find little illumination through the traditional renderings of the *jus ad bellum*, *in bello*, and *post bellum*. It is fitting that Cheyney Ryan opens the volume with a powerful exploration of the multivalent alienation felt by returning combatants. As he emphasizes, not only do soldiers experience self-alienation and self-disassociation from their pre- and postwar selves, but they also face added alienation when coming home to civil society. This suffering is intensified by the failures of public justification and responsibility for war, and it finds little balsam in the sterile and impersonal language of rights, the very language that grounds just war theory. Ryan asks us to refocus the conversation on the aftermath of war to the laments of the returning soldiers and their expressions of both grief and complaint. Among the questions Ryan poses is what it could possibly mean – not theoretically but actually – for soldiers to return home intact, spiritually and emotionally, having fought well in a good war.

Lisa Tessman continues this line of questioning in her discussion of moral injury and focuses on the discord between conventional rights-based and judicial conceptions of morality on one hand, and the lived experience of returning combatants on the other. The simple fact, as traced by the burgeoning literature on moral injury, is that soldiers frequently experience regret, shame, guilt, horror, and self-loathing, even when performing commendably under the terms and laws of combat. Tessman presents a compelling argument that we might, rightly and importantly, distinguish third-person judgments of combatants as innocent of wrongdoing from their first-person experiences of having committed some sort of moral crime, e.g., killing. This tension is at the heart both of conceptions of moral injury and the tragedy of war generally, and captures an essential element of war's tragedy, which is that would-be heroes often feel like nothing less than monstrous moral failures.

The aftermath of war for returning combatants is often its own endless war, a war within and without, lacking peace or resolution. In Chapter 3, Nancy Sherman invites reflection on how we might explore ancient Stoic philosophy to gain some purchase on this. Invoking the under-appreciated writings of Seneca on considerations of moral injury, Sherman suggests there is untapped wisdom regarding self-empathy and self-forgiveness that might rely on the same divergence examined by Lisa Tessman between first- and third-person perspectives. Inversely, perhaps, Sherman argues that the more soldiers can view themselves from the third-person viewpoint, the more they might be able to find some moral solace. It's noteworthy that this has become a practice with struggling combat veterans, who write letters to themselves from the vantage of a benevolent supervisor.

The significance of emotions, as related to the beginning and ending of wars, applies not only to combatants but also to those who would initiate wars. Similarly, the tensions between the legalistic, medieval understanding of the *jus in bello* and *jus ad bellum* create difficulties. In Chapter 4, Nir Eisikovits provides a rich phenomenology of humiliation and the ways in which this germinates into political instability, violence, and reactionism. Wars, as waged by those in power, are inevitably personal, in some sense born of very basic feelings of displacement. One need only look at the recent posturing between Russian President Vladimir Putin and NATO to see that claims for respect by those putatively disregarded are germane to war's beginning. And while the *jus ad bellum* formally denies emotional-psychological claims, such claims seem to take the reins often. This chapter pushes us to consider further and more deeply how we might navigate the "paleolithic" elements that inevitably impact group relations and the origins of war.

Shifting from personal to more structural concerns, in Chapter 5, Colleen Murphy probes (and problematizes) the very concept of war's end and the difficulties inherent in defining the *post bellum* as such. Drawing parallels between theories of transitional justice and the just war tradition, Murphy highlights their shared aims at the creation of just peace and political stability in the face of nonideal circumstances. However, both are also confronted by the failure to reckon with the indeterminate nature of such achievements as realized, in the all-too messy real world, and the degree to which imbalances of power between victors and vanquished muddy the very definition of postwar justice. Together, these complications reveal the overly idealized ambitions of theories of just war and transitional justice. In response, Murphy proposes that we establish "moral minimums," floors that constitute the duty for *enough* justice, as opposed to *aspirational* or full-fledged justice.

In a similar vein, in Chapter 6, Alex Bellamy explores the structural dimension of endless wars with a focus on why we have seen them proliferate in the past few decades. With rich historical analysis, Bellamy debunks the popular view that the increase in protracted wars is the byproduct of American interventionism and efforts to impose a global, liberal hegemony. As Bellamy illustrates, the historical facts simply belie such explanations. Rather, the increase of intractable conflicts arises from several causes, particularly the spread of intrastate – as opposed to interstate – war on the one hand, and often opposing third-party interventions on the other. Civil wars have long been recognized as particularly enduring, a quality exacerbated by the ways that these conflicts bleed beyond borders into both neighboring and remote states. Given this diagnostic account, Bellamy suggests leaders need, among other things, to better specify the precise goals of interventions, and better calibrate those proposed goals with a realistic assessment of the means available. Ambitious aims, coupled with modest means, create a recipe for protracted war, as exemplified in U.S. interventions in Afghanistan and Iraq.

In Chapters 7 and 8 respectively, David Rodin and Darrel Moellendorf further problematize the traditional just war framework by examining the *jus ad bellum* proportionality demand, i.e., the claim that wars may be waged only when the foreseeable harms are outweighed by the foreseeable values achieved. Protracted wars make such assessments exceedingly difficult, and these difficulties have yet to be sufficiently examined. Rodin discusses two salient problems that arise from the relationship between proportionality calculations and (extended) time. First, similar to any cost-benefit analysis for long-term investments, indeterminacy and optimism biases make any forecast that looks beyond the immediate future tenuous, if not suspect. The distant future is subject to unforeseeable changes, and our tendency is to ignore those prospects and/or assume the best. Second, long-term investments, in war as in finance, confront the fact of diminished value over time. Put simply, just as x fungible dollars invested today is worth more than x fungible dollars received ten years from now, the values received at the end of a protracted war may be far less valuable than the upfront investment. This future discounting, for Rodin, applies to both "blood and treasure," financial investments as well as human costs. Without addressing concerns about indeterminacy and diminishing returns, proportionality calculations prove impossible.

Moellendorf engages directly with the problem of how or whether sunk costs should inform the decision to end a war. This is also a problem of proportionality assessments through time. If the *jus ad bellum* proportionality requirement is to be workable then we must presume there is some limit to the harms done in war – what Moellendorf calls the proportionality budget – in the sense that exceeding this limit would render the war unjustified on the grounds that the harms its causes are disproportionate to the good it seeks. The problem is what to do if, despite our best projections, a war turns out to be more costly than anticipated and we exhaust the proportionality budget prior to achieving the valuable ends. In this case, are we morally obligated to end the war on the grounds that all harms caused thereafter are unjustified? Or may we discount the sunk costs and run new proportionality calculations only weighing future projected costs against the valuable ends sought? To help make progress with this dilemma, Moellendorf draws a distinction between two ways of conceiving of proportionality. In one view, the values at stake in proportionality calculations are conceived impersonally in the manner of commensurable costs and benefits. In the other view, the values at stake in proportionality calculations are composed of incommensurable values, some of which involve wrongs done to persons who are worthy of respect. Wrongs done to persons are disvalues that are incommensurable with costs such as mere expenditures of labor and money. For Moellendorf, the latter view of proportionality is less amenable to discounting sunk costs than the former and would tend to support ending wars when the initial proportionality budget has been exhausted. Moreover, this view of proportionality makes it harder to justify the resort to war because it will generate a smaller proportionality budget than the former view.

Concluding the volume, both Daniel Mauer and Daniel Philpott push the conversation from diagnosing the causes of and problems involved in endless wars toward strategies for establishing robust peace in the aftermath of lasting conflict. In keeping with Bellamy and others, both authors recognize that the majority of recent wars have involved intrastate, internation struggles, which make the creation of sustainable peace particularly arduous.

Mauer targets the unique end-of-war exigencies that are endemic to contemporary wars, addressing specifically the challenges that confront military leaders, lawyers, and combatants in establishing just conduct. Mauer illustrates how the *jus in bello* offers inadequate guidance for military action in the liminal space between the cessation of open hostility and the creation of secure peace. To fill this vacuum, Mauer suggests the creation of a new military code of conduct that is carefully molded in response to the ambiguities of the transitional environment. Drawn in parallel to the values that instruct military prosecutors, this code aims to provide principled and procedural guidance for military operations that are distinct from the *jus in bello* and more amenable to the achievement of postwar peace.

In the final chapter, Philpott offers the evocative claim that we would do well to reimagine our understanding of *post bellum* operations. Beyond the rights-based, liberal conceptions that inhere in dominant forms of transitional justice, he suggests we adopt an understanding of justice as reconciliation. Drawing from the Abrahamic religious traditions, Philpott explores reconciliation as a distinct framework for peace, one in which notions of mercy and forgiveness, for example, offer a more capacious and potentially effective avenue for establishing just peace. With a highly detailed and concrete application to the case of Iraq, Philpott illuminates how political reconciliation can both complement and supplement theories of transitional justice. Punishment, for instance, is not reduced to the juridical mathematics of exacting what is due, but is understood restoratively. Crimes are not simply violations of the law, they are more centrally a breakdown of relations. Restorative punishment responds to such crimes through the lens of healing wounds and healing relations, and punishments must be tailored to this end. In this view, if the end of war is to be both peace and justice, then we must aspire to more than the oft-shortsighted claims for rights and legalistic equality, and further address the deeper fissions and diseases that beget wars and failed relationships.

Taken together, these chapters illuminate the many gaps between conventional thinking on the ethics and laws of war, and the ways in which our understanding of war must adapt to the current realities, lest endless wars continue to devastate. Our hope is that this volume might reorient the conversation to better suit the demands of the lived experience of enduring war: sociologically, politically, militarily, and personally. Perhaps the weight of war can be better measured on scales that match its gravity, in metrics calibrated by emotion, institution, and godlike powers of destruction.

CHAPTER 1

The Lament of the Demobilized

Cheyney Ryan

You sad-faced men, people and sons of Rome,
By uproar sever'd, like a flight of fowl
Scatter'd by winds and high tempestuous gusts,

O, let me teach you how to knit again
This scatter'd corn into one mutual sheaf,
These broken limbs again into one body.

—Shakespeare, *Titus Andronicus* (Act V, Scene 3)

When Justification Comes to Grief

In his *Philosophy of Right*, Kant acknowledges that "The determination of what constitutes right in war is the most difficult problem of the right of nations and international law."[1] He proposes this constraint: war must be waged "according to such principles as render it always possible to pass out of that natural condition of the states in their external relation to each other and to enter into a condition of right."[2] That "natural condition" is one of war, which Kant identifies not just with open conflict but as much if not more with armed hostility between states inducing a constant state of fear, and which states are obliged to put behind them. He specifies this constraint on war in "On Perpetual Peace" thus: "No nation at war with another shall permit such acts of war as shall make *mutual trust* impossible during some future time of peace [emphasis mine]." He gives several examples of acts thus prohibited, such as assassination, poisoning, and breach of surrender. But they all reflect the same assumption that mutual trust *after* a war requires maintaining trust *during* a war. "Some level of trust in the enemy's way of thinking must

[1] Immanuel Kant, *The Science of Right*, trans. W. Hastie (A and D Publishing, 2018 [1790]), part II: Public Right, section 57: Right During War.
[2] Ibid.

be preserved even in the midst of war, for otherwise no peace can ever be concluded."[3]

Kant's thinking presumes a larger conception of what war and peace are all about. It assumes that the aim of war is to establish peace with the adversary; or more accurately, to *re*establish a peace that has been broken. He contrasts this with wars of "extermination," which aim at the annihilation of the enemy. And it assumes that the peace to be sought is what he terms "permanent peace," rather than a "faux-peace" that is just a temporary truce until the war starts again. It removes the causes of war, which include the reasons for threatening war as well as for actually waging it. It presumably does this by establishing the kind of trust between parties that allows for resolving disputes through discussion rather than conflict. Thus, he writes, "No treaty of peace that tacitly reserves issues for a future war shall be held valid. For if this were the case, it would be a mere truce, a suspension of hostilities, not peace, which means the end of all hostilities, so much so that even to modify it by 'perpetual' smacks of pleonasm [redundancy]."[4] The points are linked insofar as a war that aimed at *less* than "permanent peace" would be a war whose aim was simply more war.

"Who Takes Us Home Again?"

> A hundred thousand million mites we go.
> Wheeling and tacking o'er the eternal plain,
> Some black with death – and some are white with woe.
> Who sent us forth? Who takes us home again?
>
> – Charles Hamilton Sorley (1915)[5]

This bears on reestablishing peace with *another* community. What about reestablishing peace within one's *own* community? This is the problem raised in the speech from *Titus Andronicus*, above. It has special salience for a republican vision like Kant's, which saw a distinctive feature of republics to be their capacity to be at peace with themselves, in contrast to other polities riven by discord; so living in a republic was an education in peace-making that its citizens could then impart to other communities. War rends the social fabric. It disrupts lives, severs connections, strains the bonds of community, so that afterward the populace must be "knit together again." And

[3] Immanuel Kant, *Perpetual Peace and Other Essays on Politics, History, and Morals*, trans. Ted Humphrey (Indianapolis, IN: Hackett Publishing, 1983), section I, article 6.

[4] Ibid., section I, article 1.

[5] Charles Hamilton Sorley, "A Hundred Thousand Million Mites We Go," 1915, https://poets.org/poem/hundred-thousand-million-mites-we-go.

this peace-making begins with those who *fought* the war – those "sad-faced men" who now return to become part of the community again.

As I construe this republican vision, the problem of welcoming back the soldier begins with how they were sent to war in the first place. This is implicit in its conception of why wars must be publicly declared. Traditionally, the requirement that a nation be forthright about its reasons for war was meant principally for its adversaries, and since the monarch had sole authority over such matters a declaration was really a proclamation informing others what the monarch had in mind. By contrast, the republican doctrine of popular sovereignty saw war as an act of the whole community that was only legitimate if done with the full consent of the community. A declaration of war was, as it were, a performative act in which the community as a whole spoke and by which the community as a whole entrusted itself to a common project in which each citizen was accountable to the others for participating in that project. Republicanism termed this mutual accountability "fraternity." Then what if this was done fraudulently, or not at all? What happened then to the social fabric?

This was a lesson of the Vietnam War. There was never a declaration of war, partly because America's leaders could never agree on why it was being fought. So they improvised, fabricating various equally vacuous justifications, often inconsistent with each other. This impacted the prosecution of the war amid the decline of popular support and the rise of outright resistance, especially within military ranks. The leading antiwar voice before he was assassinated, Dr. Martin Luther King, Jr., stressed the breach of trust with ordinary soldiers. Along with all the other hardships, duplicity about the war's purpose had "added cynicism to the process of death."[6] And it impacted the peace afterward. Other American wars had cost more in lives and treasure. But none so shattered the community's bonds of trust. In time, this brokenness came to be identified with one "sad-faced" soldier – the Vietnam veteran. His personal struggles to return to normal life came to symbolize the country's larger struggle to "knit itself together again."

But all of this assumes that soldiers *do* return to become ordinary citizens. By this, I mean it assumes that soldiering is not some permanent occupation that sets them apart from everyone else. Yet this is what soldiering has been for much of history, and what it was in Europe prior to the republican model. The British military was typical. Its soldiers were so much a class apart that many did not come from Britain or even speak English. When the Colonists condemned them as a "standing army" essential to the "permanent

[6] Martin Luther King, "Beyond Vietnam," speech delivered April 4, 1967, https://bit.ly/3zgpTlY.

military establishments" at the heart of British society, they were invoking the contrast between this model and their own citizen-based model.

Here is what they had in mind: The older model was an *army of professionals*. For it, there was no problem of soldiers returning home because the military constituted its own *type* of home distinguished by its own way of acting and its own way of thinking – or more accurately, *non*-thinking. Its spirit was one of subservience, famously captured in Frederick the Great's remark in his *Art of Commanding an Army*: "If my soldiers began to think, not one would remain in the ranks."[7] Discipline was a matter of mindlessness, instilled by a mixture of bribery, fear, and alcohol. Its hierarchical nature meant that relations within the military were monologic insofar as some ordered and others simply obeyed.

The newer model – or ideal – was what Thomas Paine termed an *army of principles*.[8] For it, soldiering was a voluntary commitment assumed by some as part of the sacrifices assumed by all citizens in times of war. Its egalitarian nature meant that a defining feature of its relations was their *dialogic* character. In his influential pamphlet, "Easy Plan for Discipline of a Militia" (1775), Timothy Pickering insisted that "Soldiers are *reasonable* beings, as such they are to be treated."[9] Benjamin Franklin put it in terms of the citizen-soldier's right to challenge what he was told:

> It has been for some time a generally received opinion, that a military man is not to inquire whether a war be just or unjust; he is to execute his orders. All princes who are disposed to become tyrants must probably approve of this opinion, and be willing to establish it; but is it not a dangerous one? Since, on that principle, if the tyrant commands his army to attack and destroy not only an unoffending neighbor nation, but even his own subjects, the army is bound to obey. A Negro slave in our colonies, being commanded by his master to rob and murder a neighbor, or do any other immoral act, may refuse, and the magistrate will protect him in his refusal. The slavery, then, of a soldier is worse than that of a Negro![10]

George Washington famously remarked that "When we assumed the soldier, we did not lay aside the citizen."[11] This imposed a twofold

[7] Bevin Alexander, *How Wars Are Won* (New York: Crown Publishing, 2002), p. 238.

[8] Thomas Paine, "Agrarian Justice," in *The Writings of Thomas Paine, Volume III. 1791–1804* (New York and London: G. P. Putnam's Sons, 1895), p. 191.

[9] Timothy Pickering, *An Easy Plan of Discipline for a Militia* (Salem, MA: Samuel and Ebenezer Hall, 1775), part I, chapter 1: "Of the Arms and Accoutrements of a Soldier," https://bit.ly/3zJ4VoA.

[10] Benjamin Franklin, "On the Criminal Laws and the Practice of Privateering," letter to Benjamin Vaughan, March 14, 1785, in *The Works of Dr. Benjamin Franklin* (Nurnberg and New York: N. Hickman, 1835), 290.

[11] George Washington, "Address to the New York Provincial Congress," June 26, 1775, https://founders.archives.gov/documents/Washington/03-01-02-0019.

burden on the reasons given to citizens for soldiering. They must justify the citizen's leaving home, given the sacrifices that soldiering could involve, and hence must be ones that the citizen recognizes as valid in that regard. We might term this the *prospective* burden on reasons. And they allow for the citizen's return home, given the sacrifices their soldiering has involved. The reasons must be ones that the citizen can live with – and in so doing live with what they have experienced. We might term this the *retrospective* burden on reasons. Both were seen as constraints on irresponsible war-making. Citizens would refuse to become soldiers to participate in unjust wars, given the costs it would involve; former soldiers would refuse to endorse unjust wars, given the costs involved.

The problem of coming home was important to the Revolutionary generation and later. The "Headless Horseman" in Washington Irving's "The Legend of Sleepy Hollow" was the ghost of a Hessian mercenary who symbolized both the professional soldiers' inability to think and their fate to wander forever with no place to call home. But debate persisted throughout the Revolution and after on whether an army of principles was really possible. It involved a host of issues. The problem for us is that of coming home: once citizens have been called to war, is returning home truly possible? Or are the "sad-faced" soldiers that return destined to remain forever among society's "broken limbs"?

Soldier Pacifism

"I am the man who has known affliction"

– Lamentations 3:1

The question of a soldier's coming home is as old as *The Odyssey*, but it only became a focus of cultural concern as related to the common ordinary grunt with World War I. Wars had been fought for millennia, but this was the first one in which common ordinary soldiers gained a voice about their experiences, due mainly to the spread of literacy. The upshot was a *literature of disillusionment* as exemplified by novels like Erich Maria Remarque's *All Quiet on the Western Front*. A central focus was the disconnection between the dominant ideologies of war and what the first-hand experience of war was really like; in the extreme, some doubted the ability of language to capture it at all. The literature of disillusionment involved a deep skepticism of war, but not a flat-out rejection of it. Rather, it insisted that the fate of the returning soldier was a heretofore hidden cost of war that needed to be included in a full assessment of the endeavor.

The sharpest expression of this disillusionment was what I call the literature of "soldier pacifism." This is my term for a tradition of antiwar thinking that emerges from those who have experienced war and concluded that knowing what war is really like leads to a commitment against all war. Pacifism is often regarded as the height of naïveté, oblivious to war as it really is. This type of pacifism reverses this accusation, claiming it is *pro*-war thinking that does not know what war is really like. So it is a pacifism from the ground up, or the inside out. Tolstoy, the leading pacifist of the nineteenth century, was such a figure. The title of this chapter is from a major pacifist text of the twentieth century, Vera Brittain's *Testament of Youth* (1933).[12] Brittain came to her views as a frontline nurse in World War I, who lost her brother, her fiancé, and many of her friends to that war. Her phrase means to capture the claim of soldier pacifism that the experience of war is such that the "sad-faced" men who fight it can never be fully welcomed back, and that this constitutes an indictment of *all* war.

My aim in what follows is to reflect on this soldier-pacifist tradition, in part for the larger issues it raises of how we approach the topic of war. "Just war theory," as it is now understood, approaches matters from the standpoint of the observer, via the impersonal discourse of individual rights. This reflects its close relation to concerns about the laws of war. The soldier-pacifist tradition approaches matters from the standpoint of the participant – by which it does *not* mean officers or the professionally committed, but the common ordinary grunt – via the personal discourse of individual experience. A strength of this approach is that it reminds us that those who actually fight wars – soldiers – are not abstractions. But the appeal to personal experience obviously raises a great many philosophical issues about the force of such appeals, given how differently war can be experienced and how differently those experiences can be interpreted. So a concern of mine will be how to understand what the soldier-pacifist means by "knowing war" and the kind of conclusions they hope to draw from it.

A starting point is why this critique might construe itself as a "lament." To begin with, a lament is an expression of grief. Like Antigone, Brittain mourns a lost brother; both suggest that there is an intimate relation between knowing war and grieving war. And here, it is grief *at* the grief of the "sad-faced" returning soldiers. They are isolated by their grief, entombed in their grief, still at war with society in their grief (Hobbes defines the "state of war" as one in which men have only "a great deal

[12] Vera Brittain, *Testament of Youth* (London: Orion, 2014 [1933]).

of grief" in the company with one another[13]). Moreover, a lament is a complaint. It is similar in that regard to a legal objection, but its logic is quite different from the legalistic (hence it is different from philosophical arguments modeled on the legalistic). It does not aim to establish claims about an endeavor so much as reorient our responses to it. I shall note how the soldier-pacifist often conceives of themselves as a type of prophet or as speaking with a prophetic voice. In the Hebrew Bible, that voice often takes the form of lamentation (as with the prophet Jeremiah). Its aim, writes Walter Brueggemann in *The Prophetic Imagination*, is to nourish an "alternative consciousness" to the dominant ideologies of the day.[14] So we may think of the lament of the demobilized as bearing witness to the experience of war in the hope of nourishing a fundamentally different response to war. What this means and whether it succeeds can only be determined by exploring the position further.

Saigon, 1969

> They say that war is hell, but I say it's the foyer to hell. I say coming home is hell [because] hell ain't got no coordinates. Hell is no place at all, so when you're there, you're nowhere – you're lost. (Tyler Boudreau[15])

A guiding thought here is that soldiers are not abstractions. Let me say a word about how I came to this issue. Every young man of my generation faced the question of serving in the Vietnam War. The result in my case was spending a brief time in Saigon in 1969. Upon becoming draft eligible, my plan was to do two years of alternative military service that a friend of mine, Bert Bigelow (himself a prominent soldier-pacifist), offered to arrange with a medical project run by the American Friends Service Committee (AFSC) in Vietnam. They had run two projects during the war, both of them in Quan Ni province. One was making artificial limbs for amputees, the other was a pre-kindergarten school for refugee children. But this meant going to Vietnam to arrange it personally. My efforts, like the Vietnam War itself, were ultimately pointless. An AFSC project director had been captured during the Tet Offensive and, while she was eventually released, the projects were still on hold. More to the point, the draft lottery was held at exactly

[13] Thomas Hobbes, *Leviathan* (Indianapolis, IN: Hackett Publishing, 1994 [1651]), part I: Of Man, chapter 13: "Of the Natural Condition of Mankind as Concerning Their Felicity and Misery."

[14] Walter Brueggemann, *The Prophetic Imagination* (Minneapolis, MN: Augsburg Fortress, 2001), p. 3.

[15] Tyler Boudreau, *Packing Inferno: The Unmaking of a Marine* (Port Townsend, WA: Feral House 2008), p. 5.

the same time and my number left me basically exempt. So instead of going to Quan Ni province, I went to graduate school.

Following the Tet Offensive, Saigon itself was an uneventful place, bustling with American soldiers whose average age was the same as mine (21) though they seemed at once both younger and older. Many were younger (Americans killed in the war were as young as 15) but they seemed older because their fates were so uncertain. How many of the young faces I saw eventually returned to shattered lives, or never returned at all?

I didn't have much dialogue with anyone at the time. But ever since, my thinking about war has been shaped by the thought of dialogue with those young soldiers. One question it raises is the one just noted about rights-oriented discourse versus the discourse of personal experience. The issue is not one of rejecting rights-oriented discourse (and just war theory framed in such terms); rather, it is one of questioning how well that discourse captures the full reality of war and the depth of what is wrong with it. Consider this remark by Camilo Ernesto Mejía, a former Army Staff Sergeant and prominent Iraq war dissenter:

> It can be claimed that a particular war is justified politically, or that it has the support of the international community and the blessing of international law. But these arguments can never convey the images, the sounds, the smells, or anything that remotely depicts the full horror of war. Escaping those arguments is the irreversible damage war always inflicts upon humanity, and upon everything worth living on earth.[16]

The claim is that the experience of war attunes us to what "escapes" the received justifications of war and what constitutes the "full horror" of war, a horror insulting not our individual identity of personhood but our shared identity of humanity. But how should we construe the latter?

The question relates to another topic relevant here, the question of moral injury and how to understand it. It is typically characterized as an assault on the soldier's conscience, but insignificant attention is given to the meaning of "conscience." It is often described in cognitivist-type terms, as involving the violation of a soldier's "deeply held moral beliefs and expectations" contradicting their "belief system and worldview."[17] I do not doubt that the young soldiers I encountered in Saigon had consciences capable of being injured, but translating this into talk of their

[16] Camilo Mejía, *Road from ar Ramadi: The Private Rebellion of Staff Sergeant Mejia: An Iraq War Memoir* (New York: New Press, 2007), p. 299.

[17] Brett T. Litz et al., "Moral Injury and Moral Repair in War Veterans: A Preliminary Model and Intervention Strategy," *Clinical Psychology Review*, 29, no. 8 (2009), 695–706.

"belief system and world view" strikes me as minimizing the problem as experienced. How should we speak of the violation of their humanity? Here is another quote from Camilo Mejía:

> When I opened fire that day, I violated that law and desecrated the most sacred sanctuary of my being. As I observed that young man through the sight of my rifle, I was staring at a point of no return, the very Rubicon of my life, and I crossed it. My moral injury is the pain I inflicted upon the very core of my being when I took something I could never give back.[18]

What strikes me in the literature of soldier pacifism is how often the horrors of war are spoken of in terms of the "sacred" – or rather, the violation of the sacred, though without any suggestion that such talk is only intelligible with a prior theological framework. Rather, this is what the experience of war reveals – that some things are so sacred, and some acts are so sacrilegious, that they constitute a challenge to our standard discourses of justification. This fits with the traditional prophetic understanding of lament as a cry against the violation of the sacred by those who had lost any appreciation for it. But the question remains of how that cry should be understood.

Ultimately, how we talk about war will reflect who we take ourselves to be talking to. In a revealing 2002 article, "The Triumph of Just War Theory (and the Dangers of Success)," Michael Walzer characterized the audience for just war theory by way of contrast with what he termed "doctrines of radical suspicion," which would include the type of pacifism that interests me. He wrote of such doctrines, "This is the radicalism of people who do not expect to exercise power," whereas "By contrast, just war theory, even when it demands a strong critique of particular acts of war, is the doctrine of people who do expect to exercise power."[19] This strikes me as right – up to a point. As it is now practiced, just war theory approaches war from the standpoint of the policymaker, who if not someone that "exercises power" imagines themselves speaking to those that "exercise power." Whether this presumes a limited conception of "power," and an exaggerated conception of the just war theorist's impact, are matters for another time. By contrast, the assumption of my approach – which I call *critical war theory* – is that the audience for reflection on war begins with those who actually fight the wars, meaning the young men I encountered in Saigon and expected to be joining had my draft number not been #321. Theirs is a power that begins

[18] Camilo Mejia, "Healing Moral Injury: A Lifelong Journey," *Fellowship*, 76 (2011), 26.
[19] Michael Walzer, "The Triumph of Just War Theory (and the Dangers of Success)," *Social Research*, 69, no. 4 (2002), 935.

with a radical suspicion of what Vietnam vet W.D. Ehrhart has called the "pigeon-breasted fantasies" of leaders.

A Meditation

> Why didn't they want him to speak? Why didn't they want him to be seen?
> (Dalton Trumbo, *Johnny Got His Gun*)

The young man, a soldier, lies in a hospital bed. It is dark, so silent he can only feel his heartbeat. He is alone, or at least he can sense no one present. He thinks back on his experiences as a youth, joyful ones but also ones of the transition; his father dying, as his family surrounds his bed. What he remembers most are the faces: the face of his first friend, the face of his first love, the faces of both joy and mourning. He wrestles to weave together his memories in ways that constitute a single narrative, a coherent whole.

His thoughts grow curious about himself. Slowly, the self-realization begins. He is lying in bed from the injuries of war – whose full extent slowly dawns on him.

It is dark because he's lost his eyes, ripped from their sockets by an explosion. It is silent because he has also lost his ears – in fact, he no longer has a face, just a jellied mass. But he cannot touch his face, or any part of his body, because his arms have been amputated; he cannot shift his weight because his legs have been blown off. He senses no one present because he has lost the means of sense: no ears to hear others, no arms to hold another, no eyes to look into the face of others.

He is a brain in a vat. In his thoughts, he is like a child stuffed back into the womb. "Jesus, I'm in an awful mess," he thinks.

"Let us suppose that we are dreaming," Descartes writes in his *Meditations,* "and that all these particulars – namely, the opening of the eyes, the motion of the head, the forth-putting of the hands – are mere illusions; and that we really possess neither an entire body nor hands such as we see."[20] The soldier wonders if he is dreaming. Then he wonders how in his entombed state he could tell the difference. "He was in an awful mess if he couldn't even tell whether he was awake or asleep." It should be a simple matter. "It's asking very little for a man just to want to be able to prove that he's awake."

[20] Rene Descartes, *Meditations on First Philosophy* (Indianapolis, IN: Hackett Publishing, 1993 [1641]), Meditation I: "Of the Things of Which We May Doubt."

This blurring of reality and dream is not unlike the experience of war itself. Here is Vietnam vet Ron Kovic: "He heard a small girl moaning now. She was shot through the stomach and bleeding out of the rear end. He felt crazy and weak as he stood there staring at them with the rest of the men, staring down onto the floor like it was a nightmare, like it was some kind of dream and it really wasn't happening."[21] The specter here is one of madness, "craziness" in Kovic's words, which Descartes likened to falling "into very deep waters" where "I can neither make certain of setting my feet on the bottom, nor can I swim and so support myself on the surface."[22] For our young soldier, this "inability to tell dreams from thoughts" made him "nothing and less than nothing. It robbed him of the only thing that distinguished a normal person from a crazy man." He worries that his uncertainties will ultimately "rob" him of "any respect for his own thoughts and that was the worst thing that could happen to anybody."[23]

So how can this soldier – he is a guy named Joe – ward off this fate and think his way back to existence? The first step is already happening, of awakening to his full predicament. It is as if every malady of war has been compressed into his person: he is a refugee, a displaced person, a victim of a kind of torture, a prisoner of war, as well as the living abstraction that former soldiers become.

The next step is questioning. "Oh why the hell did you ever get into this mess anyhow?" He enters into a retrospective self-reflection on what he has done. He says it is a "duty" he owes himself. But it is also part of reassembling himself, piecing himself back together psychically in the hope of enduring what he now faces by trying to construct a narrative of how he ended up where he is. Vietnam vet Tim O'Brien writes eloquently of how stories can play this role, how they are especially needed "for those late night hours when you can't remember how you got from where you were to where you are."[24] Joe's whole life now is such a midnight hour.

The final step is challenging what has happened, of moving from grief to grievance. He reflects on how, as the sum of all war's maladies, he could be a fitting monument to war. Not a monument of stone to console the grieving, but a monument of flesh and blood to war's bodily reality to awaken others to the full grief that war brings. Almost like a carnival display: "See that red gash? That was his face girls. Here girls touch it don't be afraid. Bend down and kiss it. You'll have to wipe your lips afterward because

[21] Ron Kovic, *Born on the Fourth of July* (Brooklyn, NY: Akashic Books, 2005), p. 200.

[22] Descartes, *Meditation*, Meditation II: "Of the Nature of the Human Mind, and That It Is More Easily Known than the Body."

[23] Kovic, *Born on the Fourth of July*, p. 200.

[24] Tim O'Brien, *The Things They Carried* (Boston and New York: Houghton Mifflin, 1980), p. 38.

they will have a strange rotten stuff on them but that's all right. He would put a sign over it that said: he is war, concentrated into one place – and they will never forget it as long as they live." His role would be a prophetic one, alerting people to the meaning of what they do, his bodily mass a kind of resurrection. "And then suddenly he saw. He had a vision of himself as a new kind of Christ as a man who carries within himself all the seeds of a new order of things. He was the new messiah of the battlefields saying to people as I am so shall you be."[25]

The story of Joe is from the major American antiwar novel of the twentieth century, Dalton Trumbo's *Johnny Got His Gun*.[26] Young men like me first learned of it in the Vietnam War era from Ron Kovic's memoir, *Born on the Fourth of July*. Kovic, a disabled Marine turned antiwar leader, wrote "This novel came into my life when I desperately needed to know that I was not alone. I wanted to not feel so lost, so *angry* and *afraid*, and I wanted to reconnect with a part of me that I was so afraid of."[27] It finds a new audience in every American conflict. Cindy Sheehan's son, Casey, was killed in Iraq in April 2004. She has written of how the novel was central to her own "quest for truth and for peace" that led her to become a tireless opponent of that war, and later war generally.[28]

In the talk of becoming a "messiah of the battlefields," Trumbo's novel especially evokes the prophetic role that survivors of war sometimes envision for themselves. It is as if there is something about the experience that compels them to this role. Here is Kovic: "The one gift I was given in that war was an awakening. I became a messenger, a living symbol, an example, a man who learned that love and forgiveness are more powerful than hatred, who has learned to embrace all men and women as my brothers and sisters."[29] Their fate in war has imposed this special burden. Writes Kovic, "It is we who must cry out for a world without war," for "Nothing is more important than for us who have experienced war to share its awful truth."

The problem posed by what I call soldier pacifism is how we should understand this "awful truth" that they feel compelled to share. I approach it with two thoughts. We may assume that the "truth" in question here is not an esoteric one, i.e., it is available to anyone who has experienced war. But at the same time, it needs articulation to be fully present to people,

[25] Dalton Trumbo, *Johnny Got His Gun* (New York: Citadel, 2007), p. 148.

[26] Ibid.

[27] Ron Kovic, "Introduction (1990)," in Trumbo, *Johny Got His Gun*, p. xv.

[28] Cindy Sheehan, *Peace Mom: A Mother's Journey Through Heartache to Activism* (New York: Atria Books, 2010).

[29] Kovic, *Born on the Fourth of July*, p. 24.

including those who have experienced war. It needs, in Kovic's words, a "messenger" whose principal concern is to find the proper language in which the experience can be cast. Some accounts of the "awful truth" are actually quite straightforward. Here is Vietnam vet Camillo Bica: "In fact, no one knows the sacrilege of war better than we who must fight it and then have to live with the memories of what we have done and what we have become."[30] So here we encounter the appeal to the sacred/sacrilege. And talk of such notions seems to fit the kind of truth just identified. Traditionally, being sacrilegious has been seen as something that everyone is capable of acknowledging. It is available to everyone, but it still needs to be presented in the right way, especially if ideologies seek to obscure it – in the case of war, by claiming that individuals, actions, and even places are sanctified by its carnage. Hence its critics have conceived of themselves as messengers, prophets – those whose role is to awaken us to the sacrilegious in our lives. But how can we unpack such talk through the experiences recounted?

War, the Face, the Face-to-Face

Every war has its signature injury. A history of war could be written by its distinctive maimings. Anglo-Saxon skirmishes gave us blunt trauma by clubs, maces, and battle hammers; the American and French Revolutions, amputations. The American Civil War added gangrene and sepsis, World War I lung injuries and trench foot. Brain injuries have been a prominent feature of recent conflicts. Iraq and Afghanistan have witnessed unprecedented numbers of wounds to the genitals.

But injuries to the face have a special status. World War I's trench warfare proved diabolically conducive to facial injuries to the point that combat could be experienced as an ongoing encounter with damaged faces. "We see men living with their skulls blown open; we see soldiers run with their two feet cut off, they stagger on their splintered stumps into the next shell-hole. We see men without mouths, without jaws, without faces; we find one man who has held the artery of his arm in his teeth for two hours in order not to bleed to death. The sun goes down, night comes, the shells whine, life is at an end."[31] But this was always hidden. It constituted a "hidden history" of that war in one historian's words.[32] Photos of facial injuries were banned. Limbless veterans

[30] Mac Bica, cited in Rita Brock and Gabriella Lettini, *Soul Repair: Recovering from Moral Injury after War* (Boston: Beacon Press, 2013), p. 21.

[31] Erich Maria Remarque, *All Quiet on the Western Front* (New York: Ballantine Books, 1987), p. 134.

[32] Caroline Alexander, "Faces of War," *Smithsonian Magazine*, February 2007.

had many support groups but the facially disfigured were left to themselves. In Sidcup, England, site of the first hospital devoted to facial wounds, some park benches were painted blue as a warning that the man sitting on one would be distressful to view. Some proposed that entire villages be designated where the "maimed and shattered" could live by themselves. But this did not address the problem of soldiers facing themselves. Hospitals banned mirrors because men who glimpsed themselves often collapsed in shock.[33] A recent estimate is that over 38,000 facial injuries occurred in Iraq. One of the more famous stories to emerge from that conflict, Brian Van Reet's "Big Two-Hearted Hunting Creek," speaks to facial injuries in that conflict.[34]

The face is the site of our most deeply human capacities like grief, shame, and tears – and anger. The face-to-face is the site of our most deeply human interactions with each other. What does it mean to lose your face?

The Face as Site of Reckoning

> We are the ones who have to live
> with the memory that we were the instruments
> of your pigeon-breasted fantasies.
> We are inextricable accomplices in this travesty of dreams:
> but we are not alone.
>
> – W. D. Ehrhart, "A Relative Thing"[35]

Let me begin with the face of anger. Anger and retribution are persistent themes in the literature of disillusionment. A "favorite fantasy" of British soldiers in World War I was "visiting of violent and if possible painful death upon the complacent, patriotic, uncomprehending, fatuous civilians at home."[36] Correspondent Phillips Gibbs wrote of soldiers that "They prayed to God to get the Germans to send Zeppelins to England – to make the people know what war meant."[37] And the anger turned to rage when the war was over. Siegfried Sassoon's "Fight to a Finish" imagined

33 Olga Khazan, "Masks: The Face Transplants of World War I," *Atlantic Magazine*, August 2014.

34 In Roy Scranton and Matt Gallagher (eds.), *Fire and Forget: Short Stories from the Long War* (Boston: Da Capo Press, 2013).

35 W. D. Ehrhart, "A Relative Thing," http://vietnamwar.lib.umb.edu/experience/docs/poem_WDEhrhart.html.

36 Paul Fussell, *The Great War and Modern Memory* (New York: Oxford University Press, 1975), pp. 83–84.

37 Philip Gibbs, *Now It Can Be Told* (Scotts Valley, CA: CreateSpace Independent Publishing Platform, 2014) p. 195.

soldiers marching through London in a welcome-home parade, fixing their bayonets and attacking the crowd. Poet Charles Hamilton Sorely put it bluntly: "I should like so much to kill whoever was primarily responsible for the war."[38] A half-century later the "pigeon-breasted fantasies" of those responsible for the Vietnam War were the target of former Marine Sergeant and Purple Heart winner W.D. Ehrhart in the poem just cited. "My day of reckoning is upon me. Yours will come" – these are the words of Tomas Young who was wounded and paralyzed in Iraq five days after arriving. His powerful "A Message from a Dying Veteran" is an imagined dialogue with the architects of that war, written on behalf of "those who will spend their lives in unending pain and grief" and confronting them to "face what you have done to me and to many, many others who deserved to live."[39]

Young's dialogue is an imagined one because there is in fact no dialogue of the sort that animated the army of principles model. Invoking some ideas of G.A. Cohen, the breakdown of accountability can be understood as a breakdown of the face-to-face justifications identified with fraternity.[40] In his critique of Rawls, Cohen invites us to contrast justifications that are "blandly impersonal" with interpersonal – or face-to-face – ones. The first is a *third*-person type justification, meaning it is a justification from no one in particular for people in general. If successful, it establishes an abstract moral obligation on everyone to do some X. The other is a *second*-person type justification, meaning it is a justification from someone in particular to someone in particular. If successful, it establishes a concrete normative claim of one person on another that they do some Y. And the conditions for their success are quite different. The soundness of a third-person justification is solely a matter of what is said, while the soundness of an interpersonal face-to-face justification is also a matter of the relation between the parties, of who is speaking and who is being spoken to.

The problem with third-person faceless justifications is not just whether you agree with them, abstractly, but whether you can care about them, concretely, which is a matter of your caring-type relations with the community of which you are a part. Implicit in the orientation of soldier pacifism is that anything like soldiering, involving killing and

[38] Quoted in Jon Silkin, *Out of Battle: The Poetry of the Great War* (London: Palgrave Macmillan, 1998), p. 52.

[39] This letter can be found, along with a video of Tomas Young reading it, at *Democracy Now*, "Exclusive: Tomas Young Reads in Full His Letter to Bush & Cheney, 'A Message From a Dying Veteran,'" March 21, 2003, https://bit.ly/3vWzDAN.

[40] Gerald Allan Cohen, *Rescuing Justice and Equality* (Cambridge, MA: Harvard University Press, 2008).

dying, should only be done for reasons that one cares about, which partly involves a community of mutual caring. But despite all the rhetoric, this is precisely what does not exist – nor can it exist in the conditions of modern war. The upshot is that, despite the rhetoric, the relation of society to soldiers is inevitably one of turning away. Indeed, I think another "hidden history" of modern war would concern the many different strategies employed for this turning away of the face. The most recent one is today's mantra of "thanking the troops" without the slightest interest in who is being thanked and what they are being thanked for. I wonder about the reaction of the young men I encountered in Saigon if I had begun by "thanking them for their service." Maybe anger: there is a moment in *Johnny Got His Gun* where Joe's senses that a group of officers has arrived to pin a medal on him. The whole situation is absurd. Given his physical condition, they can't really know whether he fought for their country or the other side. He is just a faceless "soldier," but they are careful to cover his face while there so they cannot see the full meaning of what they are "thanking" him for. Joe is increasingly angry, though they cannot know this from his hidden face.

The Face as Site of Revelation

The face is a locus of identity, especially as fashioned by our trials and tribulations, so to look in the soldier's face is to look for what soldiering means. But what if there is an inscrutability to that experience that thwarts not only full comprehension by others but by the soldiers themselves? This is the struggle for a language to articulate the experience, driven by a frustration at how the received discourses fall short.

Here is an episode from Kovic's *Born on the Fourth of July*. It is typical of this literature in describing an incident of true dilemma, i.e., one where terrible things are done, yet not in violation of the rules of engagement, raising the question of the claim that such rules have on the experiences in question.

The soldiers of his unit fire into a village where they honestly believe the enemy is firing at them.[41] Afterward, one of them enters a hut.

> 'Oh God,' he said. 'Oh Jesus Christ.' He started to cry. 'We just shot up a bunch of kids!' The floor of the small hut was covered with them, screaming and thrashing their arms back and forth, lying in pools of blood, crying wildly, screaming again and again. They were shot in the face, in the chest,

[41] Kovic, *Born on the Fourth of July*, pp. 198–202.

> in the legs, moaning and crying. 'Oh Jesus!' he cried. There was an old man in the corner with his head blown off from his eyes up, his brains hanging out of his head like jelly. He kept looking at the strange sight, he had never seen anything like it before.

A dialogue ensues between them and their commanding officer. As others arrived,

> The men were not moving and some of them were crying now, dropping their rifles and sitting down on the wet ground. They were weeping now with their hands against their faces. 'Oh Jesus, oh God, forgive us.' The men in the hut were just sitting there crying. They could not move, and they did not listen to the lieutenant's orders. They just sat with the rain pouring down on them through the roof, crying and not moving.

Their lieutenant is honestly concerned with helping the victims, but he is frustrated by his men's reactions. He finally blurts out: "'You're men, not babies. It's all a mistake. It wasn't your fault. They got in the way. Don't you people understand – they got in the goddamn way!'"

We might take the point of the lieutenant to be that the rights of the children and older people had not been violated because the killing was accidental; he might even invoke the lessons of liability theory to establish this. But how much does this show? Part of the problem with the juridical language of rights, responsibilities, personhood, etc. is that it imagines a kind of closure, a "yes" or "no" answer, in these contexts. One wonders if it can really address the problem I raised above of living with what one has done. Kovic and others consistently speak of being "haunted" by what they have done – what kind of ethical claim "haunts us" in this way? It seems to involve the claim of the face. Kovic stresses heads blown off, brains hanging out.

Remarque's *All Quiet on the Western Front* famously stresses the *eyes* of the enemy that, in his words, "cry out," "yell" in ways that are disempowering, or in the Kovic incident debilitating. Remarque links this with engaging the enemy's identity. He writes:

> But you were only an idea to me before, an abstraction that lived in my mind and called forth its appropriate response. It was that abstraction I stabbed. But now, for the first time, I see you are a man like me. I thought of your hand-grenades, of your bayonet, of your rifle; now I see your wife and *your face* and our fellowship. Forgive me, comrade. We always see it too late. Forgive me, comrade; how could you be my enemy?[42]

[42] Erich Maria Remarque, *All Quiet on the Western Front* (New York: Everyman's Library, 2018) p. 223.

But the moral resistance here involves the inability to fully capture it in words. Kovic speaks of the "strangeness" of the experience. Camilo Mejía speaks with the same tentativeness of a similar case where he acted within his rights, but the moral problem remained:

> For weeks after the incident, my mind could not shake off the images of the young man walking, and breathing, and then down on the ground, bloody, and dead. I had also followed a lawful order, and I had not opened fire until I was convinced that he was indeed going to throw a grenade. The problem was that as I observed that young man through the sight of my rifle, when he was still alive, there was something inside me, a voice one could say, that was telling me not to squeeze the trigger. And I knew, without a shred of doubt, that I should not disobey that voice, and that if I did, there would be serious consequences to face.[43]

The "voice" here is the voice of conscience that continues long after issues of abstract justification have been resolved. All of this pertains to killing in war. It also extends to dying in war. The dead are those who never return, but the post-World War I institution of "memorials" to the dead, including "tombs to the unknown soldier," suggest that they are nevertheless still with us. Ever since, the quarrel about war has involved a quarrel about the meaning of such memorials. William Faulkner's antiwar novel, *A Fable*, recounts a Christ-like soldier who sparks a mutiny on the Western front, is executed for his resistance, but then, by a series of flukes, ends up being the "unknown soldier" buried to "honor the dead."[44] Significantly, the pro-war ideology of honoring the dead invokes the question of accountability. Beginning with such pro-war poems as John McCrae's "In Flanders Fields," killing – and ultimately dying – in war is something that citizens are called to do by the voices of those already dead. McCrae stages it as a dialogue in which "we, the dead" hold the living accountable to "Take up our quarrel with the foe." It is as if impersonal justifications are not enough, something more personal is required.[45]

The discourse of soldier pacifism reverses this. A recent striking example is Larry Heinemann's, *Paco's Story*[46], in which the tale of a broken Vietnam veteran is narrated by the ghost of a fellow soldier (ghost stories are common in both the American and Vietnamese literature of that war). Kovic speaks of his standpoint as that of a "corpse, the living dead

43 Bica, in Brock and Lettini, *Soul Repair*, p. 87.
44 William Faulkner, *A Fable* (New York: Random House, 2011).
45 John McCrae, "In Flanders Fields," www.poetryfoundation.org/poems/47380/in-flanders-fields.
46 Larry Heinemann, *Paco's Story* (New York: Farrar Straus Giroux, 1984).

man." In *Johnny Got His Gun*, Joe ridicules the idea of the dead speaking, while claiming the right to speak as one who is as close to being dead as possible. I quote at length his interior monologue about the reasons for dying in war:

> And all the guys who died all the five million or seven million or ten million who went out and died to make the world safe for democracy to make the world safe for words without meaning how did they feel about it just before they died?
>
> If the thing they were fighting for was important enough to die for then it was also important enough for them to be thinking about it in the last minutes of their lives. That stood to reason. So did all those kids die thinking of democracy and freedom and liberty and honor and the safety of the home and the stars and stripes forever?
>
> You're goddam right they didn't. They died crying in their minds like little babies. They forgot the thing they were fighting for the thing they were dying for. They thought about things a man can understand. They died yearning for the face of a friend. They died whimpering for the voice of a mother a father a wife a child. They knew what was important.[47]

I take these remarks to be addressing the status of the reasons for what happens to soldiers and the distinction between agreeing with reasons and caring about them, the assumption being that the reasons we can live with must be ones we can care about and even more – reasons that we die for must be ones that we can care about. Before, the question was killing for abstractions, now it is one of dying for abstractions, which in both cases are posed in terms of the experience of the face-to-face. If killing requires turning away from the face of the other, the suggestion here is that dying compels us to turn toward the face of another, specifically of a loved one. Or, to put it another way, the only type of meaning worth dying for is that which can be found in such face-to-face encounters. But this is what the inhumanity of modern war denies.

The Cry of Hope

> No one truly "recovers" from war. No one is ever made whole again. The best that can be hoped, I think, is to achieve a degree of benign acceptance. To that end, I strive each day to forgive and absolve myself of guilt, and to live with the wounds of war that will never heal.
>
> – Camillo Baca[48]

[47] Trumbo, *Johnny Got His Gun*, pp. 71–73.

[48] *Peace Vet*, "Personal Legacy of the War," December 17, 2011, www.svaphilosopher.com/Peersonal-Legacy-of-War.html.

From the start, the troubled story of the returning Vietnam veterans was one of grief. Their plight first gained attention with the 1971 story of Sergeant Dwight Johnson.[49] Sergeant Johnson was the first African American soldier to be awarded the Congressional Medal of Honor in Vietnam, but after two more years of distinguished service he came home increasingly dejected about his experiences and fate. He was shot and killed robbing a liquor store. This was the result of many factors. But in one of the first discussions of the "Post-Vietnam Syndrome," psychiatrist Chaim F. Shatan suggested that the inability of soldiers to truly come home was the profound consequence of "impacted grief."[50] The talk of "impacted" invokes my characterization of Joe as embodying all the maladies of war in his one person.

What does it mean to construe this as a violation of "humanity"? I want to conclude with some observations by the French philosopher Simone Weil on the language of protest. Weil was deeply critical of characterizing forms of deep oppression in the juridical language of rights. She is among those thinkers who feel that the evils of our time can only be captured in the language of the sacred/sacrilege, while recognizing the challenges to finding a place for such talk in our secular culture. Here she is writing about the degradation of labor in modern life, and how workers should understand what might be termed the "moral injury" that it involves:

> It is sacrilege to degrade labour in exactly the same sense that it is sacrilege to trample upon the Eucharist. If the workers felt this, if they felt that by being the victim they are in a certain sense the accomplice of sacrilege, their resistance would have a very different force from what is provided by the consideration of personal rights. It would not be an economic demand but an impulse from the depth of their being, fierce and desperate like that of a young girl who is being forced into a brothel; and at the same time it would be a cry of hope from the depth of their heart.[51]

There are two thoughts here that I think worthy of further reflection within the discourse of war. One is the concern that to frame the costs of war, specifically the grief it induces in the language of individual rights is ultimately to privatize that injury in a way that compounds the privatizing

[49] Jon Nordieimer, "From Dakto to Detroit: Death of a Troubled Hero," *The New York Times*, May 26, 1971, www.nytimes.com/1971/05/26/archives/from-dakto-to-detroit-death-of-a-troubled-hero-from-dakto-to.html.

[50] Chaim F. Shatan, "Post-Vietnam Syndrome," *The New York Times*, May 6, 1972, https://www.nytimes.com/1972/05/06/archives/postvietnam-syndrome.html.

[51] Simone Weil, *Selected Essays* (Eugene, OR: Wipf and Stock, 1962), p. 18.

impacted grief itself. Weil sees the language of the sacred as articulating our pains by lifting us out of them, as it were. The second thought is one that informs the lament of the demobilized especially as expressed in the voices we have considered. How do we maintain hope? This is the question that confronts all grief and I think it is a question that confronts any sober-minded reflection on the persistence of war in our time. Weil's suggestion is that any discourse that construes itself as a dialogue with the afflicted must enable the "cry of hope." And this begins by listening to its voices.

CHAPTER 2

Moral Injury and Moral Failure

*Lisa Tessman**

Introduction

Military service – particularly combat – puts service members at a high risk of having an experience of moral failure of a kind that few civilians face: the experience of having to perform an action – often killing someone or letting someone die – that one realizes is the best thing that one can do in the situation, but that nevertheless strikes one as being morally wrong or even unthinkable. Grasping the unthinkability of an action that one has to perform is an emotional experience of extreme aversion and distress. The experience may result in what is called moral injury, namely, in the words of one veteran, "the kinds of wounds we inflict on ourselves that come inherently with the wounds we inflict upon others."[1] Killing another person – especially if it is someone who is particularly vulnerable, such as a child – or not saving a person whom one loves or for whom one is responsible are among the primary actions that cause moral injury.

Moral injury is considered to be *moral* because of the real or perceived moral wrongdoing that is at its source and because it is characterized by certain moral attitudes – attitudes of holding oneself responsible or taking

* I first presented an earlier draft of this chapter at the Ethics of War and Peace Conference, which was held at the United States Military Academy, West Point, in October 2019; I also presented versions of it at the Institute for Advanced Studies in the Humanities at Binghamton University in February 2020; at the Central New York Moral Psychology Workshop in November 2020; at the Jackson Family Center for Ethics and Values, Coastal Carolina University, in February 2021; and as the Sullivan Lecture at Fordham University, in April 2021. I benefitted enormously from comments from participants at these events. I am grateful to graduate students in my Spring 2020 seminar on Moral Responsibility, who read and discussed a draft. I would also like to thank individuals who commented on earlier drafts, including: Mich Ciurria, Barrett Emerick, Laura Engel, Ami Harbin, Alex King, Katie Stockdale, and Mark Wilson. Most of all, I thank my spouse, Bat-Ami Bar On, for contributing to my thoughts in this chapter in many ways; as a veteran herself, she knew a lot about living in the aftermath of war. She died unexpectedly on November 16, 2020.

[1] Tyler Boudreau, "The Morally Injured," *The Massachusetts Review*, 52, no. ¾ (2011), 748.

oneself to have failed morally, and feelings of guilt about this failure. And moral injury is called an *injury* because it often manifests in debilitating symptoms and self-destructive behaviors that, at the more extreme end, include withdrawal, abuse of drugs and alcohol, and self-harm, including suicidal ideation and attempts at suicide.[2] There are mental health practitioners who specialize in treating moral injuries, and researchers who investigate what causes moral injury, how it can be prevented, and how the symptoms can be more effectively treated. This is all important work, and I believe that service members and veterans are owed much better treatment than what is currently available. But this is not my focus. Instead, I am interested in what service members and veterans who experience moral injury thereby express about moral failure. I want to salvage the "tragic truth" that I believe they often express – namely that some of the actions that it may be beyond our control to avoid are not just wrong but unthinkable – and that tends to get dismissed because of the discrepancy between their own attitudes of holding themselves responsible, and other people's attitudes of releasing them from responsibility.[3]

Experiences of Moral Injury

According to the model of moral injury developed by some of the leading psychiatrists and psychologists in the field, the experiences that may cause moral injury include:

> perpetrating, failing to prevent, bearing witness to, or learning about acts that transgress deeply held moral beliefs and expectations… Moral injury requires an act of transgression that severely and abruptly contradicts an individual's personal or shared expectation about the rules or the code of conduct, either during the event or at some point afterwards.[4]

[2] Sonya B. Norman and Shira Maguen, "Moral Injury," National Center for PTSD, www.ptsd.va.gov/professional/treat/cooccurring/moral_injury.asp.

[3] As Peter Marin writes about this "tragic truth": "Our great therapeutic dream in America is that the past is escapable, that suffering can be avoided, that happiness is always possible, and that insight inevitably leads to joy. But life's lessons… teach us something else again, something that is both true of, and applicable to, the experience of the vets. Try as they do to escape it, the past pursues them; the closer they come to the truth of their acts, the more troubled they are, the more apart they find themselves, and the more tragic becomes their view of life." Peter Marin, "Living in Moral Pain," in Robert Emmet Meagher and Douglas Pryer (eds.), *War and Moral Injury: A Reader* (Eugene, OR: Cascade Books, 2018 [1981]), 178.

[4] Brett Litz, et al., "Moral Injury and Moral Repair in War Veterans: A Preliminary Model and Intervention Strategy," *Clinical Psychology Review*, 29, no. 8 (2009), 700. Psychiatrists and psychologists who treat moral injury clinically and who research the causes and treatments of moral injury include Brett Litz, Shira Maguen, and William Nash. Like them, I am focused on cases in which the injury comes from one's own moral failure. However, Litz, Maguen, Nash, and others' conception

The U.S. Department of Veteran Affairs recognizes that "In the context of war, moral injuries may stem from direct participation in acts of combat, such as killing or harming others, or indirect acts, such as witnessing death or dying, failing to prevent immoral acts of others, or giving or receiving orders that are perceived as gross moral violations."[5] A series of studies led by Shira Maguen has revealed that killing in war is highly associated

of moral injury is a departure from the understanding of moral injury that was introduced by psychiatrist Jonathan Shay; see Jonathan Shay, *Achilles in Vietnam: Combat Trauma and the Undoing of Character* (New York: Simon and Schuster, 1994); and Jonathan Shay, *Odysseus in America: Combat Trauma and the Trials of Homecoming* (New York: Scribner, 2002). According to Shay, the injurious situations include those in which the injured party is injured not by perpetrating or being complicit or associated with wrongdoing, but rather by being betrayed or wronged by someone in authority. Using Achilles' betrayal by Agamemnon in the *Iliad* as his model, Shay describes moral injury as a "betrayal of what's right" by someone in authority (Shay, *Achilles in Vietnam*, p. 3). Shay himself distinguishes his definition of moral injury from that of Litz, Maguen and Nash, noting that the term "has been used in two related, but distinct, senses; differing mainly in the 'who' of moral agency. Moral injury is present when there has been (a) a betrayal of 'what's right'; (b) either by a person in legitimate authority (my definition), or by one's self – "I did it" (Litz, Maguen, Nash, et al.); (c) in a high stakes situation"; Jonathan Shay, "Moral Injury," *Psychoanalytic Psychology*, 31, no.2 (2014), 182. Shay adds that "there's not a lot we can do to eliminate the sort of Moral Injury that Nash, Maguen, and Litz have written about"; "Moral Leadership Prevents Moral Injury," in Robert Emmet Meagher and Douglas Pryer (eds.), *War and Moral Injury: A Reader* (Eugene, OR: Cascade Books, 2018 [2010]), 304. Shay believes that given his own definition, moral injury can be prevented through better military leadership, because it is betrayal on the part of the leadership that causes moral injury. He also points out that betrayal by the leadership – for instance, by giving commands that require immoral actions – is what often leads to service members' own moral failures, and thus to moral injury as defined by Litz, Maguen, Nash and others (ibid., 305). Loosely following Shay, some scholars employ a very broad understanding of moral injury; Nancy Sherman, for instance, writes that moral injury "refers to experiences of serious inner conflict arising from what one takes to be grievous moral transgressions that can overwhelm one's sense of goodness and humanity. The sense of transgression can arise from (real or apparent) transgressive commissions and omissions perpetrated by oneself or others, or from bearing witness to the intense human suffering and detritus that is a part of the grotesquerie of war and its aftermath. In some cases, the moral injury has less to do with specific (real or apparent) transgressive acts than with a generalized sense of falling short of moral and normative standards befitting good persons and good soldiers"; Nancy Sherman, *Afterwar: Healing the Moral Wounds of Our Soldiers* (New York: Oxford University Press, 2015), p. 8. See also Nancy Sherman, *The Untold War: Inside the Hearts, Minds, and Souls of Our Soldiers* (New York: W.W. Norton and Company, 2010). I am examining a narrow rather than broad range of situations that can cause moral injury. While I believe that moral injuries that result from one's own moral failure and those that result from the failures of leadership are both important, there are crucial differences in the emotions or moral attitudes that are associated with each, and they should not be lumped together. This is confirmed by Schorr et al. in a study of what veterans perceive as having been morally injurious in their military experience: "Based on our qualitative results, we posit that an important organizational framework for considering morally injurious events is the veteran's attribution of responsibility for the event. An event in which he or she feels responsible for what occurred (either through an act of omission or commission) is likely to result in different emotional responses, and thus have different implications for treatment than events in which the responsibility is attributed to others (e.g. guilt and shame versus anger and frustration)"; Yonit Schorr et al., "Sources of Moral Injury Among War Veterans: A Qualitative Evaluation," *Journal of Clinical Psychology*, 74, no. 12 (2018), 12.

5 Norman and Maguen, "Moral Injury."

with a number of different mental health and functional consequences, which together can be taken to characterize moral injury. Furthermore, the severity of the effects was found to depend on the type of killing that the veteran had engaged in; to put a rough gloss on it, those who have killed civilians exhibit more severe symptoms than those who have killed only enemy combatants, and those who have killed women, children, and/or the elderly, as well as those who have killed prisoners, suffer more than those who did not.[6] It is not simply killing, or killing of a certain kind, but also the veteran's attitude toward their own killing that matters; the attitude of "combat guilt" is strongly associated with, for instance, suicidal ideation and suicide attempts.[7] Retrospective thoughts or memories of their own acts of killing – particularly but not only forms of killing that were not justified – are often accompanied by guilt, shame, a feeling of having failed morally, or a judgment that one is a bad person. As Noah Pierce, an Iraq War veteran, wrote in his suicide note, "I am not a good person, I have done bad things. I have taken lives, now it's time to take mine."[8] The concept of moral injury is meant to capture a linked cluster of experiences (such as killing), moral emotions (such as guilt and shame), and harmful effects; this cluster differs from the cluster that characterizes the more well-known phenomenon of post-traumatic stress disorder (PTSD), which typically includes life-threatening experiences, the primary emotion of fear, and symptoms such as hypervigilance.

[6] Shira Maguen et al., "The Impact of Killing in War on Mental Health Symptoms and Related Functioning," *Journal of Traumatic Stress*, 22, no. 5 (2009), 435–43. In a study of Vietnam combat veterans, those who had killed had significantly higher rates of "PTSD symptoms, dissociation experiences, functional impairment, and violent behaviors," even after controlling for other kinds of combat exposure (ibid., 440). A later study of Vietnam veterans found that those with "killing experiences were twice as likely to report suicidal ideation as those who did not kill, even after accounting for PTSD, depression, substance use disorder diagnoses, and adjusted combat exposure"; Shira Maguen et al., "Killing in Combat May Be Independently Associated with Suicidal Ideation," *Depression and Anxiety*, 29 (2012), 921. In a similar study of Iraq War veterans, Maguen and her colleagues again found a strong relationship between killing – whether directly or indirectly – and "PTSD symptoms, alcohol abuse, anger, and relationship problems"; Shira Maguen et al., "The Impact of Reported Direct and Indirect Killing on Mental Health Symptoms in Iraq War Veterans," *Journal of Traumatic Stress*, 23 (2010), 86. A study of Gulf War veterans found that "taking a life in combat is a critical ingredient in the development of postdeployment mental health concerns"; Shira Maguen et al., "The Impact of Killing on Mental Health Symptoms in Gulf War Veterans," *Psychological Trauma*, 3, no. 1 (2011), 24. And a study of both Iraq and Afghanistan veterans found that those "with particular killing experiences involving non-combatants and/or anger or revenge are most at risk"; Shira Maguen et al., "Killing and Latent Classes of PTSD Symptoms in Iraq and Afghanistan Veterans," *Journal of Affective Disorders*, 145, no. 3 (2013), 348.

[7] Herbert Hendin and A.L. Haas, "Suicide and Guilt as Manifestations of PTSD in Vietnam Combat Veterans," *The American Journal of Psychiatry*, 148, no. 5 (1991), 586–91.

[8] Ashley Gilbertson, "The Life and Lonely Death of Noah Pierce," *The Utne Reader*, February 13, 2009. www.utne.com/politics/lonely-death-noah-pierce-ptsd-iraq-war.

Killing is not the only act that is experienced, in the context of military operations, as a moral failure. Other moral transgressions may include acts such as torturing or humiliating prisoners or detainees, destroying property, sexually assaulting civilians or fellow service members, injuring others either intentionally or accidentally, using disproportionate or unnecessary violence, acting cruelly, or ordering others to commit a variety of actions, as well as omissions such as failing to protect a fellow service member or a civilian, being negligent in a way that results in harm or death, or not intervening when witnessing someone else's immoral behavior. While all of these acts can result in a moral injury to the service member who commits the act, some cases differ from others in an important respect: in some cases, the act is fully avoidable, while in other cases, there is no way that the service member could have avoided doing what they did without doing something worse or equally bad. Many cases lie somewhere in between the avoidable and the unavoidable. I will be focusing primarily on unavoidable failures, with the understanding that it is frequently hard to know whether or how a particular act could have been avoided; often what haunts a service member or veteran is the thought that perhaps they could have done something better than they did. When an act cannot be avoided, at least not without doing something worse or equally bad, then the moral status of the act is complicated by the fact that one can say both that it was the best possible thing to do and *in that sense* was the right thing to do, while in some other sense it was still a wrongdoing.[9] But before getting to unavoidable moral wrongdoings, I will say a few words about those that are avoidable.

Consider cases in which a service member tortures or humiliates prisoners for entertainment, rapes a civilian or another service member, or uses more force or violence than is necessary for accomplishing a task. For example, think about the following story that a military chaplain told to journalist David Wood, about a soldier who:

> ...had been in a firefight and had his rifle sight trained on an insurgent, but instead of firing at the man's chest – the "center of mass" that troops are taught to aim for – he consciously lowered his weapon and shot him in

[9] Most moral philosophers would deny that there can be such a thing as an impossible-to-avoid moral wrongdoing, because they accept the principle that "ought implies can" – that is, that one cannot be morally required to do anything that one cannot do. I have argued that "ought implies can" does not apply to all moral requirements and that there indeed are impossible moral requirements and unavoidable moral failures. See Lisa Tessman, *Moral Failure: On the Impossible Demands of Morality* (New York: Oxford University Press, 2015); and Lisa Tessman, *When Doing the Right Thing Is Impossible* (New York: Oxford University Press, 2017).

> the stomach, knowing that would cause a lingering and painful death. As an army medic worked to save the man, the soldier and his squad members gathered around and watched the man die in agony.[10]

The soldier came some months later to the chaplain because he was suffering from the anguished feelings characteristic of moral injury; the soldier had realized: "'I'm not just killing because I have to, but because I want this to be a suffering,' and he knew that to be wrong. A painful violation of his own moral values."[11]

In this case, the soldier could have and should have avoided the wrongdoing. The distress he experienced afterward is entirely appropriate. In fact, it is a sign that he has at least some moral conscience. When veterans self-reflectively blame themselves for intentional, avoidable moral wrongdoings, they are right to take responsibility for their actions. Some therapeutic approaches have assumed that moral injury (like PTSD) is a matter of being plagued by *distorted* cognitions – such as the belief that one is guilty of wrongdoing when in fact one is not – and that the therapist should aim to change the distorted self-blaming attitudes. Other mental health professionals contend that this assumption about distorted cognitions is misguided, precisely because there are cases in which the veteran has committed a moral wrongdoing and *properly* grasps this. For instance, Gray, Nash and Litz point out that "individuals can and do occasionally act intentionally in ways that violate their values and standards of conduct, and... such actions – when reflected upon rationally – may give rise to significant guilt, shame and distress... In cases in which at least partial culpability is real and rational, the assignment of blame to oneself or others must also be rational, appropriate, and accurate."[12] They argue that in such cases it is a mistake for the therapist not to recognize that self-blame is warranted.[13] It is fairly clear that veterans who react to their own avoidable wrongdoings by taking responsibility are accurate when they take themselves to have failed morally, though of course they might take either too much or too little responsibility for an action for which they are only

[10] David Wood, *What Have We Done: The Moral Injury of Our Longest Wars* (New York: Little, Brown and Company, 2016), p. 21.

[11] Ibid.

[12] Matt Gray, William Nash, and Brett Litz, "When Self-Blame is Rational and Appropriate: The Limited Utility of Socratic Questioning in the Context of Moral Injury," *Cognitive and Behavioral Practice*, 24 (2017), 384.

[13] Their approach, called Adaptive Disclosure, thus assumes that: "(a) pain means hope – anguish, guilt, and shame are signs of an intact conscience and self-and-other-expectations about goodness, humanity, and justice; (b) goodness is reclaimable over the long haul; and (c) forgiveness (of self and others) and repair are possible regardless of the transgression"; ibid., 386.

partially to blame. When veterans hold themselves responsible for wrongdoings that they could have and should have avoided committing, they are likely to see their own response to themselves – self-blame – reflected in a parallel reaction from other people, who may also blame the veteran.

However, this tends not to be the case when veterans hold themselves responsible for having committed some terrible act that was at the same time the best action they could have taken in the circumstance – they will not see their reaction to themselves reflected in other people's reactions, for other people will have a tendency *not* to hold them responsible. In these sorts of cases veterans may have committed the terrible act under some degree of compulsion, or in the face of a moral dilemma in which all available options would have involved wrongdoings, or in situations in which they lacked crucial information for making a decision, or lacked sufficient control to do what would have been right. Alternatively, they may have committed the act because the only alternative would have been extreme self-sacrifice. In other words, the action that the veterans understand to have been wrongdoings are not necessarily actions that they could have avoided, at least not without doing something worse or equally bad, or not without self-sacrificing more than anyone expects them to. For instance, consider this case, related by David Wood, in which a marine probably did the best thing that he could have done in the situation:

> Nikki...was twenty-two years old when he was sent as a marine infantryman to Afghanistan, where he shot and killed a young boy. This was not uncommon in the murderous confusion of our recent wars, where farmers and mothers and young kids might seize a weapon and shape-shift in a moment into a combatant and back again to an innocent civilian, and young Americans peering into the murk would have a moment to decide: kill or not. This time, an exhausting firefight with Taliban insurgents had dragged on for hours across the superheated desert wastes and tree-lined irrigation canals of Helmand Province. Late that afternoon, Nik saw from the corner of his eye someone darting around the corner of an adobe wall, spraying bullets from an assault rifle held against his small hips. Nik swiveled his M4 carbine, tightened his finger on the trigger, and saw that it was a boy of maybe 12 or 13. Then he fired. According to the military's exacting legal principles and rules, it was a justifiable kill, even laudable, an action taken against an enemy combatant in defense of Nik himself and his fellow marines. But now Nik is back home in civilian life, where killing a child violates the bedrock moral ideals we all hold. His action that day, righteous in combat, nonetheless is... a painful violation of the simple understanding of right and wrong that he and all of us carry subconsciously through life.[14]

[14] Wood, *What Have We Done*, pp. 7–8.

In Nik's own words: "'He was just a kid. But I'm sorry, I'm trying not to get shot and I don't want any of my brothers getting hurt, so when you are put in that kind of situation…it's shitty that you have to, like…shoot him. You know it's wrong. But…you have no choice'."[15] Nik recognized that he faced a moral dilemma: he had to choose between the wrongdoing of killing a child and the wrongdoing of not protecting his fellow marines.

It is not always a moral dilemma that makes moral failure unavoidable. Sometimes the failure occurs because of mistakes due to unavoidable ignorance – for instance, ignorance about whether an approaching individual is an innocent civilian or an insurgent. David Wood's story about two marines, Doss and Canty, illustrates this:

> There was an Afghan kid who used to come around the marines' outpost, basking in their attention. One day he kept trying to get them to go outside, wanting to show them something. Doss and Canty and some others finally went with him into an adjacent field toward a tree line and discovered a desiccated corpse, a man the marines had shot a while back… The man had been acting suspiciously and wouldn't respond to the warning shots the marines had fired in accordance with the ROE [rules of engagement], and so he was shot and killed. Now his body had been ravaged by dogs. It turned out that the corpse was the boy's father, who was deaf and mute and couldn't hear or respond to warning shots, the boy explained, and that's why he appeared to ignore the marines and why he was shot and killed and was now lying dead in a field.[16]

One of the marines, Doss, whose own father had long been ill and who died two years after Doss' return from Afghanistan, began to have nightmares in which it was his own father whose body was lying half eaten away in the field.

Although there are particular incidents like these in which something that is easily granted to be morally horrifying – such as killing a child, or killing a vulnerable civilian – is unavoidable, the type of killing that is most clearly sanctioned in war – that is, the killing of an enemy combatant – may also be experienced as a moral wrongdoing, and killings of this type are unavoidable as long as war is unavoidable.[17] Killing enemy

[15] Ibid., pp. 13–14.

[16] Ibid., p. 140.

[17] As a study on moral injury by Schorr et al. notes: "Injuring and killing the enemy in battle is central to the core mission of war. Service members are trained to kill and depart for war with the expectation that this is what they are setting out to do. And yet, a number of veterans…described being changed by taking these actions that violated or, at the very least, challenged their previously held moral beliefs"; Schorr et al., "Sources of Moral Injury Among War Veterans," 6.

combatants might at the same time achieve one morally good aim – if the aim of the mission is itself morally justifiable, carrying out the mission necessitates killing, and the killing is conducted in accordance with humanitarian laws – while violating other moral values. As Yonit Schorr et al. put it:

> an individual may be able to justify his actions given the context within which the event occurred (e.g. self-defense or defense of others) and may even acknowledge having honored one set of morals by this action (defense of country, following orders, protecting his men), and yet still experience considerable distress about how those actions conflict with a separate set of morals (e.g. do not kill, protect the weak).[18]

Thus the simple fact of moral value pluralism – namely, the existence of multiple, irreducibly different moral values – and the inevitability of conflict between the moral requirements that these different values entail is enough to make moral failure unavoidable. It is not always possible for service members to simultaneously fulfill all of the nonnegotiable moral requirements that they face.[19] Because war requires killing, and because each human life that is lost is of the kind of irreplaceable value that we tend to take ourselves to be morally required not to sacrifice, war regularly makes moral failure inescapable, even for those who always perform the morally best action that is possible in the circumstance.[20]

In the case of *avoidable* wrongdoing, other people, including psychotherapists, may readily find it appropriate for veterans to see themselves as having failed morally. What about in the case of unavoidable wrongdoing? Veterans often still feel an anguished sense of responsibility for what they take to have been their own moral failures, even when they recognize that those failures were unavoidable. However, when the failure was unavoidable, or could have been avoided only by doing something worse or equally bad, or through extreme self-sacrifice, other people do not tend to hold them responsible despite the fact that the veterans hold themselves

[18] Ibid., 6.

[19] For my full discussion of moral value pluralism and its relation to unavoidable moral failure, and for my distinction between negotiable and nonnegotiable moral requirements, see Tessman, *Moral Failure* and *When Doing the Right Thing is Impossible*.

[20] For an important discussion of the relationship between moral value pluralism and veterans' expressions of moral failure, see Tine Molendijk, "Toward an Interdisciplinary Conceptualization of Moral Injury: From Unequivocal Guilt and Anger to Moral Conflict and Disorientation," *New Ideas in Psychology*, 51 (2018), 1–8. See also Tine Molendijk, Eric- Hans Kramer, and Désirée Verweij, "Moral Aspects of 'Moral injury': Analyzing Conceptualizations on the Role of Morality in Military Trauma," *Journal of Military Ethics*, 17 (2018), 36–53.

responsible.[21] This is an interesting mismatch of reactions. It might seem that the mismatch should be resolved by assuring veterans that they are not responsible for their unavoidable actions,[22] but I argue that this denies something important that their distress reveals.

I believe that both veterans' tendencies to hold themselves responsible and other people's reluctance to hold them responsible get something right.[23] The veterans may be right that they are responsible in the sense of having failed to meet a binding moral requirement, even when the requirement was impossible to meet. This makes their tendency to hold themselves responsible a *fitting* response because it correctly represents their own action as a wrongdoing. However, others may at the same time be right that the veterans are not responsible in the sense of being blameworthy for having violated shared normative expectations that are fair. To blame others for certain failures, including those that are unavoidable, is inappropriate for a specific reason: it reflects normative expectations that cannot be shared expectations, because one cannot reasonably expect others to accept that certain actions – including those that are impossible to perform – are required. But we may (rightly) have expectations of ourselves that are not shared expectations and that do not have to be reasonable. My claim depends on acknowledging both the *fittingness* of taking oneself to be responsible for an unavoidable failure, and the *unfairness* of the normative expectation that one avoid the unavoidable. I will suggest

[21] The mental health professionals (such as Gray, Nash, and Litz) who deny that the anguished emotions characteristic of moral injury are necessarily "distorted cognitions," and who advocate accepting rather than challenging the veteran's sense of responsibility for the action, seem to leave open the possibility that actions that are unavoidable in the context of war may still be appropriately conceived of by the veteran as moral failures. However, Gray, Nash, and Litz do argue in support of their approach by emphasizing that "moral injury can and often does result from instances of intentional perpetration" (Nash and Litz, "When Self-Blame is Rational and Appropriate," 383), so it seems that they have *avoidable* actions in mind, and do not commit one way or the other to the appropriateness of taking responsibility for *unavoidable* actions. See also Litz et al., "Moral Injury," 695–706; and William Nash and Brett Litz, "Moral Injury: A Mechanism for War-Related Psychological Trauma in Military Family Members," *Clinical Child and Family Psychology Review*, 16, no. 4 (2013), 365–75.

[22] Nancy Sherman's chapter in this volume aims to find resources in the reactions of others to help those suffering from moral injury. The approach I take here explains how self-guilt for unavoidable actions can be fitting and that acknowledgment by others of this self-guilt can be an important part of the recovery process.

[23] In some ways, this is similar to Susan Wolf's position, though she is focused on cases of moral luck rather than specifically on cases of unavoidable moral wrongdoing, so she has no reason to say, for instance, that the agent is right in taking themself to have committed a wrongdoing; rather, her point is that there is a virtue that has to do with taking responsibility for one's actions when, due to luck, they have a bad outcome. See Susan Wolf, "The Moral of Moral Luck," *Philosophic Exchange*, 31, no. 1 (2000), 1–16. Thanks to Barrett Emerick for pointing out the connection with Wolf's thinking.

that the phenomenon of moral injury should be addressed in a way that both acknowledges taking responsibility as a fitting response and recognizes the value of having shared normative expectations that are fair.

Reactive Attitudes

Work on what are called "reactive attitudes" offers a helpful framework for understanding moral responsibilities. Work in this tradition takes our sentiments – and specifically the sentiments that we experience in reaction to how other people regard or disregard us and how we regard or disregard others – and our associated practices of holding each other, and ourselves, responsible to be the basis for our responsibilities. I accept this overall approach but find one aspect of it to be problematic: existing work on the topic has for the most part denied that we can be responsible for unavoidable moral failure, and it has pushed to the side the reactive attitudes that are indicative of this kind of responsibility.[24] I think that there is a sense in which we can be responsible for unavoidable failures, and failures that would require great self-sacrifice to avoid, even when others cannot rightly hold us responsible for them, and that a case can be made for this by rethinking the implications of the reactive attitudes that are experienced as moral injury.

Reactive attitudes include both *self-reactive attitudes*, which are the attitudes a person has in response to the quality of their *own* regard for others, and *interpersonal reactive attitudes*, which are reactive attitudes that people have in response to the quality of *another* person's regard for ourselves or for others. In typical cases in which a moral injury results from an unavoidable action, there is a discrepancy between the reaction that a

[24] Similarly, work on moral injury has pushed the self-reactive attitudes to the side by suggesting that veterans are wrong to hold themselves responsible for many of their combat-related actions. As Camillo Bica noted in one of the earliest discussions of moral injury, "veterans are advised to ignore what has occurred, to 'deresponsibilitize,' i.e. to neutralize their feelings by accepting the 'naturalness' of their behavior on the battlefield…"; Camillo C. Bica, "A Therapeutic Application of Philosophy: The Moral Casualties of War: Understanding the Experience," *International Journal of Applied Philosophy*, 13, no. 1 (1999), 82. An exception is Krista Thomason who discusses the guilt that former child soldiers experience even while other people (such as mental health professionals) assure them that they are not responsible for what they did. In this case, what leads people to exempt the child from moral responsibility is the very fact that by virtue of being a child, they are not considered a full moral agent. Furthermore, even if the child is considered to be a moral agent, the fact that they were coerced to become a soldier is thought to excuse them from responsibility. Nevertheless, Thomason argues that the child's guilt is both intelligible and appropriate in the sense of being morally valuable. Krista Thomason, "Guilt and Child Soldiers," *Ethical Theory and Moral Practice*, 19, no. 1 (2016), 115–27. Thanks to Katie Stockdale for directing me to Thomason's paper.

moral agent (in this case, a service member or veteran) has to themself (the self-reactive attitude) and the reactions of other people to the moral agent (the interpersonal reactive attitudes). The typical self-reactive attitude in these sorts of cases is to feel distress in the form of guilt, shame, and self-blame, while the typical interpersonal reactive attitude is not any kind of blaming response; the distressed self-reactive response seems to reflect the way that it matters to the moral agent that they have morally failed (even if unavoidably), while the absence of an interpersonal blaming response seems to reflect an understanding that the moral agent had no morally better option, or that choosing the morally better option would have been supererogatory (that is, good but not required), so no shared normative expectation for them to have acted differently could be a fair expectation. There is a lack of symmetry between the typical self-reactive attitude and the interpersonal reactive attitude; I believe that *both* types of reactions can be appropriate, and thus that it is right for there to be an asymmetry. My aim is to make sense of the asymmetry by distinguishing between different ways in which a reactive attitude can be appropriate.[25]

I will begin with a brief review of the way that Peter Strawson characterized the reactive attitudes when he first introduced the concept in 1962.[26] Taking interpersonal reactive attitudes (as opposed to self-reactive attitudes) as the paradigm case, he notes that we have natural reactions – in the form of sentiments, or attitudes – to how other people regard and treat

[25] This is a kind of self-other asymmetry that Michael Slote has argued is absent from at least a certain version of virtue ethics. See Michael Slote, "Morality and Self-Other Asymmetry," *The Journal of Philosophy*, LXXXI, no. 4 (1984), 179–92; and *From Morality to Virtue* (New York: Oxford University Press, 1992). I disagree with Slote about this because I deny that the self-other asymmetry is due to a vice related to an improper balance between self-regarding and other-regarding traits. I don't think that the asymmetry is necessarily due to any shortcoming. Rather, I argue that we may *rightly* hold ourselves responsible for things that no one else can hold us responsible for. See Lisa Tessman, "The Virtues of Reactive Attitudes," *The Journal of Value Inquiry*, 55 (2021), 437–56. Thanks to Nancy Sherman for making the connection between Slote's work on this topic and my claims in this chapter.

[26] P.F. Strawson, "Freedom and Resentment," *Proceedings of the British Academy*, 48 (1962), 187–211. Strawson develops the concept in the context of a discussion of determinism – a context that I will leave to the side. Strawson argues that accepting "the truth of a general thesis of determinism" (ibid., 195) would have no effect on our practices of responsibility – it would not stop us from having reactive attitudes that express how and in what way other people's regard for us (and for others) matters to us. It would not lead us to excuse people from responsibility for their actions (where we tend to excuse when a bad action turns out not to have been based on any underlying ill will), and it would not lead us to exempt all people from counting as responsible agents (as, for instance, we would exempt certain categories of people, such as young children) and thus to see people as "an object of social policy…as a subject for what…might be called treatment…as something…to be managed or handled or cured or trained" (ibid., 194). Instead, regardless of the truth of determinism, the

us, that is, to their attitudes toward us.[27] They are called *reactive* attitudes because they are attitudes in reaction to attitudes, from the perspective of those engaged as participants in interpersonal relationships. Reactive attitudes contain a mixture of affective and cognitive elements; they lie somewhere between how we feel in response to what we take someone's regard to be and how we judge someone's regard. To capture this mixture, I refer to reactive attitudes as a kind of *evaluative response*. The reactive attitudes present someone's regard for others as having certain evaluative features; for instance, my resentment might present the quality of someone's regard for me as insulting or slighting, and my gratitude might present the quality of someone's regard for me as kind or benevolent. It is because it matters to us how other people regard and treat us that, when we engage with them, we tend to react in positive or negative ways to their quality of regard for us; these reactions reflect the ways in which their regard matters to us. Strawson emphasizes "how much we actually mind, how much it matters to us, whether the actions of other people – and particularly of *some* other people – reflect attitudes towards us of goodwill, affection, or esteem on the one hand or contempt, indifference, or malevolence on the other."[28] Our interpersonal reactive attitudes, in reflecting the fact that it matters to us how others regard us, serve as a "certain sort of demand for inter-personal regard."[29] The fact that regard matters gives us a constitutive reason to have a particular reactive attitude, namely one that makes a demand for regard. This is what connects *mattering* with *holding responsible*.[30] Our reactive attitudes "rest on, and reflect, an expectation of, and demand for, the manifestation of a certain degree of goodwill or regard

fact that we are committed to having interpersonal relationships in which other people's quality of will matters to us means that while it would be logically possible to adopt this "objective attitude" towards others rather than responding to them as a participant in an interpersonal relationship (i.e. responding with the reactive attitudes), it is "practically inconceivable" (ibid., 197) that we would do so; our valuing of interpersonal relationships precludes it.

27 Theorists in the Strawsonian tradition have refined what Strawson calls "quality of will" into several different aspects of our wills. For instance, David Shoemaker argues that we sometimes have a reactive attitude in response to someone's "quality of character," sometimes in response to their "quality of judgment" and sometimes in response to their "quality of regard"; David Shoemaker, *Responsibility from the Margins* (New York: Oxford University Press, 2015). Given this division, my focus is on quality of regard.

28 Strawson, "Freedom and Resentment," 191.

29 Ibid., 201.

30 This characterization borrows from Pamela Hieronymi's account, in which she distinguishes constitutive reasons from extrinsic reasons; Pamela Hieronymi, "The Wrong Kind of Reason," *The Journal of Philosophy*, CII (2005), 437–57. See also Hieronymi's notion of "responsibility as mattering," which she presents in several places, including in her 2020 unpublished manuscript "Introduction" to *Minds That Matter*, available at: https://hieronymi.humspace.ucla.edu/in-progress/.

on the part of other human beings towards ourselves; or at least on the expectation of, and demand for, an absence of the manifestation of active ill will or indifferent disregard."[31]

Given our "human commitment to participation in ordinary interpersonal relationships," and given that in such relationships it matters to us how others regard us, it would be "practically inconceivable" for us to give up our reactive attitudes and thus our practices of holding others responsible.[32] In this conception, our attitudes and our practices of holding responsible are antecedent to our responsibilities; they are the (or at least a) source of our responsibilities. As Gary Watson explains: "It is not that we hold people responsible because they *are* responsible; rather, the idea (our idea) that we are responsible is to be understood by the practice, which itself is not a matter of holding some propositions to be true, but of expressing our concerns and demands about our treatment of one another."[33] This is an important point because it means that the answer to the question of whether or not we are responsible for certain moral failures (including those that are unavoidable) is to be found in our reactive attitudes – in our practices of holding responsible that reflect what matters to us, not in some independently existing fact about what responsibilities we have or what is morally required of us.

However, not every reactive attitude establishes what we are "really" responsible for, because some of our reactive attitudes are simply inappropriate responses that we should dismiss rather than take to have any normative implications for us. We need to consider what the criteria are for a reactive attitude to be appropriate. If self-reactive attitudes of feeling responsible for moral failure are inappropriate (as we might think that they are either because failure was unavoidable, or because our only alternative would have been supererogatory), then perhaps they are not indicative of our having failed to meet any real responsibilities.

What makes a reactive attitude appropriate or inappropriate? To begin with, the reactive attitudes are appropriate only when they "fit" what they are reactions to. For a reactive attitude to fit its object, we must correctly apprehend the features of the object – whatever it is we are reacting to – and our reaction must be evaluative, such that it reflects the way in which the object of our reaction really does or does not matter to us; because a reactive attitude is an evaluative response, what it reflects when it fits

31 Strawson, "Freedom and Resentment," 200.
32 Ibid., 197.
33 Gary Watson, *Agency and Answerability: Selected Essays* (Oxford: Clarendon Press, 2004), p. 222.

its object are the evaluative features of the object. Thus, I could have an unfitting reaction either by making a mistake about what the object is, or by being somehow mistaken about what really matters. The first kind of mistake is straightforward: if I take you to have ignored me and I resent you for ignoring me, then in order for my resentment to be appropriate, I must be correct that you have ignored me; if it turns out that you weren't ignoring me – you simply didn't see me – then my resentment will be inappropriate. Because the same action could be performed with quite different underlying attitudes, the response to an action will be an inappropriate – unfitting – reactive attitude whenever one makes a wrong assumption about the underlying attitude. Take Strawson's example:

> If someone treads on my hand accidentally, while trying to help me, the pain may be no less acute than if he treads on it in contemptuous disregard of my existence or with a malevolent wish to injure me. But I shall generally feel in the second case a kind and degree of resentment that I shall not feel in the first.[34]

Once someone's intention in performing an action becomes apparent, it may (though in some cases may not) serve to either justify or excuse what would otherwise have appeared to be a wrongdoing. We would think someone *justified* in treading on my hand if they did it in order to squash the spider on my hand; we would *excuse* them for treading on my hand if they did not even see that my hand was there, because they were focused on swatting the spider off of my head; either way, they had no ill will, so if ill will is what matters – which, in Strawson's view, it is – then a reactive attitude of resentment would be inappropriate.[35]

One might thus conclude that any reactive attitude that serves to hold someone responsible for performing an *unavoidable* action – no matter how terrible the action is – is inappropriate. It may seem that whenever an agent could not have avoided a particular action, it is inappropriate to hold them responsible, simply because they have exercised no ill will. Their actions will seem to be either justified or excused. For instance, if a soldier kills an enemy combatant because doing so was necessary for meeting an overriding moral requirement to pursue the aims of a war (that was itself morally justified), their action may (assuming it meets various other

[34] Strawson, "Freedom and Resentment," 191.

[35] As J.L. Austin said of justifications and excuses, when we have a justification for our action, "we accept responsibility but deny that it was bad" and when we have an excuse for our action, "we admit that it was bad but don't accept full, or even any, responsibility"; J.L. Austin, "A Plea for Excuses," *Proceedings of the Aristotelian Society*, New Series, 57, no. 1 (1956–57), 2.

criteria, such as being in accordance with the rules of engagement) count as *justified.* If a soldier kills an innocent civilian whom they unavoidably mistook to be posing a threat, their action may be *excused.*[36] Thus interpersonal reactive attitudes that *forbear* from holding veterans responsible for unavoidable actions – even when the actions are terrible – seem to get it right; they seem to be fitting.

Furthermore, it is not just for entirely *unavoidable* actions that it may be appropriate to forbear from holding other people responsible by accepting justifications or excuses. When an action is unavoidable, we recognize that it was not performed out of ill will or lack of regard of any kind. But often what matters to us about other people is that they regard others with what we take to be *enough* good will – we do not necessarily expect that their concern for others will always override all other considerations, including a healthy regard for themselves. If someone chooses not to sacrifice their life in order to avoid perpetrating some terrible act, such as killing a child, others may be inclined to excuse their choice or see it as justified. When people do not blame veterans for acting in self-defense in such situations, it reflects a sense that it is unreasonable to expect others to make certain self-sacrifices. It matters to us that someone acts with concern toward others, but our normative expectations in interpersonal relationships allow the participants to also exercise some level of concern for themselves. Our typical interpersonal reactive attitudes serve to constitute certain levels of self-sacrifice as supererogatory. Just as we tend not to hold other people responsible when they fail to do the impossible, in many cases we also do not hold them responsible when they choose not to make extreme self-sacrifices. There will, however, be a limit to this. There are acts so terrible – atrocities – that sacrificing one's own life to avoid being a perpetrator may be morally required rather than supererogatory. And service members – by virtue of accepting their role in the military – are already committed to making sacrifices that in other contexts would be seen as supererogatory.

[36] In addition to justifications and excuses, there are also *exemptions* from responsibility. We *exempt* someone from responsibility if they are simply incapable of being responsible, for instance, because they are a young child, or an adult who is disabled from agency either permanently (such as someone who suffers from dementia) or temporarily. What counts as having one's own agency compromised for reasons beyond one's control can be difficult to determine. For instance, consider a service member who is driven by sudden and uncontrollable rage to take cruel revenge on an enemy combatant who has just killed their closest friend. Some people might tend to say the service member was in a state in which they could not be responsible for their actions; other people may be reluctant to grant this, and the service member themself may rightly resist being treated as a nonagent.

This tendency to release other people from responsibility stands in contrast to what research into moral injury has shown: that service members or veterans frequently hold themselves responsible for unavoidable actions and actions that would have entailed what other people take to be supererogatory self-sacrifice. It turns out that it is really just our *interpersonal* reactive attitudes – our reactions to *other* people – that follow a pattern of not holding people responsible for unavoidable or difficult-to-avoid failures. In the same sorts of situations, the typical *self*-reactive attitude is a feeling of responsibility for the action. However, theorists have tended to dismiss self-reactive attitudes as misguided whenever they diverge from interpersonal reactive attitudes. It seems that the guilt, shame, and self-blame that are characteristic of moral injury would, in this view, be deemed inappropriate self-reactive attitudes, and thus attitudes that do not tell us anything about what we "really" are responsible for. Is this right? Should we just dismiss the self-reactive attitude that consists of a feeling of responsibility for what are understood as unavoidable moral failures or failures to do what others take to be supererogatory?

I will suggest that we should not dismiss self-reactive attitudes just because they do not match what we agree are fitting interpersonal reactive attitudes. Rather, I will argue that there are differences between the criteria by which self-reactive attitudes should be judged to be fitting and the criteria by which interpersonal reactive attitudes are judged to be fitting, because self- reactive attitudes may fittingly be either a subjective evaluative response or an evaluative response that reflects shared normative expectations, while interpersonal reactive attitudes can only be fitting when they reflect what can reasonably be expected to be shared values or shared normative expectations. However, before examining the differences in what makes self-reactive and interpersonal reactive attitudes *fitting*, I will turn in the next section to another possible way in which one might object that the self-reactive attitudes that are typical in cases of moral injury are inappropriate, for "inappropriate" does not always mean "unfitting."

Right and Wrong Kinds of Reasons

It is possible for a reactive attitude to be fitting but to be inappropriate in some other way; for instance, a reactive attitude may be called inappropriate in the sense that experiencing it is harmful or inexpedient. As Justin D'Arms and Daniel Jacobson have pointed out, the term "appropriate" tends to be applied in ways that conflate fittingness with other concerns.[37] This "conflation problem"[38] is evident, I believe, in the charge that it is inappropriate to hold oneself responsible for unavoidable moral failures or failures to do what others take to be supererogatory. Distinguishing fittingness from other concerns will allow us to see that the self-reactive attitudes that are typical in cases of moral injury may be fitting despite there being reasons of other kinds – reasons based on other kinds of concerns – not to hold oneself responsible for unavoidable moral failures or failures to do what others take to be supererogatory. Reasons that bear on the question of what an object's evaluative features are are the "right kind of reasons" for counting the object as actually having those evaluative features; other kinds of reasons that weigh in favor of or against having an evaluative response are the "wrong kind of reasons" because they are merely *extrinsic* reasons – that is, they do not bear directly on the question of whether or not an object has those evaluative features – instead, they may be moral or prudential reasons for or against having a particular evaluative response.[39] An attitude fits its object when the reasons for calling it

[37] Justin D'Arms and Daniel Jacobson, "The Moralistic Fallacy: On the 'Appropriateness' of Emotions," *Philosophy and Phenomenological Research*, 61, no. 1 (2000), 65–90. They argue that "an emotion can be fitting despite being wrong to feel" for either moral or prudential reasons; ibid., 65. As they put it elsewhere, "to call a response 'appropriate' is vague praise… only certain good reasons for or against having a response bear on the associated evaluative judgment"; Justin D'Arms and Daniel Jacobson, "Sentiment and Value," *Ethics*, 110, no. 4 (2000), 731. What is important about this for them is that it poses a problem for theories such as (neo)sentimentalism that rely on what they call the "response dependency thesis," namely the thesis that "to think that X has some evaluative property Φ is to think it appropriate to feel F in response to X"; ibid., 729. The problem is that it may be inappropriate to feel F in response to X for reasons that have nothing to do with whether or not X has the evaluative property Φ, so it is impossible to know whether or not X has the evaluative property Φ simply on the basis of whether or not it is "appropriate" to feel F.

[38] Ibid, 732. This problem also came to be called the "wrong kind of reasons problem." See Wlodek Rabinowicz and Toni Rønnow-Rasmussen, "The Strike of the Demon: On Fitting Pro-Attitudes and Value," *Ethics*, 114, no. 3 (2004), 391–423.

[39] It is Pamela Hieronymi who introduces the terminology of "constitutive reasons" (to designate what had been called "right kinds of reason") and "extrinsic reasons" (to designate what had been called "wrong kinds of reasons"), and it is she who discusses the "wrong kind of reasons problem" in terms of how a reason bears on a question. She argues that we should reject the definition of a reason as "a consideration that counts in favor of an action or attitude" (because this definition does not allow us to distinguish right from wrong kinds of reasons) and instead "think of a reason as a consideration that bears on a question," for then we can distinguish between kinds of reasons "by distinguishing between kinds of questions on which a consideration can bear." (Hieronymi, "The Wrong Kind of Reason," 437–38.) A consideration becomes a reason when it stands in a certain relation to the question on which it bears. That is, it becomes a reason because of the relation "between the question on which the consideration bears and the attitudes of which it counts in favor" (ibid., 438). Hieronymi suggests that we consider the relation between "*settling a question* and *forming or revising an attitude*" (ibid., 447).

appropriate are reasons of the right kind, reasons that bear on the question of whether or not the object has the evaluative features that the attitude presents it as having.

D'Arms and Jacobson argue that we commit what they call the "moralistic fallacy" when we infer that an emotion is unfitting just because it would be morally wrong to feel that emotion.[40] Similarly, we might fallaciously infer that an emotion is unfitting because it would be harmful, inexpedient, or otherwise imprudent to feel the emotion. To illustrate the problematic conflation of fittingness with other criteria, D'Arms and Jacobson offer several examples: for instance, a joke may have the evaluative feature of being funny, or a colleague's promotion may have the evaluative feature of being enviable. The emotion of amusement presents its object – a joke – as being funny; the emotion of envy presents its object – a colleague's promotion – as being enviable. There are circumstances in which it would be morally wrong or inexpedient to feel amusement or envy – such as when a funny joke is racist or when you are depending on your enviably promoted colleague for an affirmative vote on your own personnel case – but the moral wrongness or imprudence of feeling a certain emotion may have no bearing on whether the emotion is fitting. A tendency to moralize everything makes us reluctant to see that an emotion that is morally wrong to feel could still be fitting; because we know that it is wrong to be amused by a racist joke, we do not like to admit that such a joke could still be funny. We may resist the point less when we are asked to distinguish between the imprudence of feeling an emotion and the fittingness of the emotion. For instance, it is quite plausible that while there are prudential reasons not to

When the relation is a *constitutive* relation, then settling the question "amounts to forming the attitude" (ibid.). For instance, settling the question of "whether *p*" amounts to forming the attitude of belief (that is, the belief that *p*), so "reasons that (are taken to) bear on whether *p* [are] the 'constitutive' reasons for believing *p*" (ibid.). We believe *p* by virtue of settling the question of whether *p*, and we settle the question when we are convinced by the reasons that (are taken to) bear on the question of "whether *p*." So too with evaluative responses such as resenting. Settling the question (in the affirmative) of whether you slighted me (given that slighting matters to me) amounts to forming the attitude of (for instance) resentment, so reasons that are taken to bear on whether you slighted me (given that slighting matters to me) are constitutive reasons for resenting you. In contrast to *constitutive reasons* for an attitude, when one develops an attitude for *extrinsic reasons*, finding the reasons convincing is not constitutive of having the attitude.

The fact that resenting you will teach you a lesson is (even given that I value teaching you a lesson) an extrinsic reason for resenting you; by being convinced that resenting you will teach you a lesson, I do not thereby form the attitude of resentment. Settling the question of whether resenting you will teach you a lesson does not have a constitutive relation to having the attitude of resentment. It merely motivates me (if I want to teach you a lesson) to try to cultivate the attitude of resentment. Importantly, it tells me nothing about whether your quality of regard has certain evaluative features, namely whether it is a form of disregard, a slighting.

40 D'Arms and Jacobson, "The Moralistic Fallacy."

fear the wolf that is coming toward you, since the wolf will smell your fear, you nevertheless cannot infer from the fact that it would be unwise to fear the wolf that the wolf is not fearsome; it *is* fearsome, and fear is a fitting, and in that sense appropriate, response to the wolf.[41]

Let us consider, then, what concerns might lead someone to say that it is inappropriate for a person to hold themself responsible for an unavoidable action by having a self-reactive attitude that presents the action as a wrongdoing. The first objection might be based on a concern about the consequences of holding oneself responsible in this way: when veterans hold themselves responsible for unavoidable but terrible actions, as they commonly do in cases of moral injury, this has seriously harmful effects on them. Recall that moral injury often manifests not just in feelings of guilt or shame, but also in self-destructive behaviors such as drug and alcohol abuse, self-harm, and (attempts at) suicide. Hence, one might correctly conclude that there is a reason for veterans *not* to hold themselves responsible even for terrible actions that they committed in the course of their service: doing so is self-destructive. This reason, however, would be an *extrinsic* reason: it does not bear on the question of whether or not their actions have certain evaluative features, such as the feature of being a wrongdoing, or being unthinkable, in ways that matter to them. The self-reactive attitude of guilt could fittingly reflect the way that it matters to Nik, for instance, that he killed a child; his feeling that "it's wrong"[42] might fittingly present that killing as a wrongdoing for which he holds himself responsible. The good reason that he has to *not* experience this self-reactive attitude (namely, whatever harm or distress the feelings or associated behavior may cause him) does not bear on the question of whether or not the killing has the evaluative features that the self-reactive attitude presents it as having.

Another concern that might underlie the charge that it is inappropriate for someone to hold themself responsible for an unavoidable action is a concern about fairness or a sense that veterans do not deserve to be the targets of the self-reactive attitudes by which they hold themselves responsible. This objection, too, conflates questions of fit with other kinds of concerns. Whether or not someone *deserves* blame is a consideration that bears on the question of whether they should be punished, but not on the question of whether their reactive attitude (which may indeed be a blaming attitude) correctly reflects the evaluative features of the object, such as the feature of being a wrongdoing. What the evaluative features of the object are depends on what rightly

41 Ibid., 87.

42 Wood, *What Have We Done*, pp. 13–14.

matters to the person whose reactive attitude it is. As Pamela Hieronymi suggests, "mattering, not meriting, is... central" to moral responsibility.[43] What she calls the "merited-consequences conception" of moral responsibility – according to which a reactive attitude is fitting if it is the reaction that someone's moral failure deserves or merits, the way one can be deserving of punishment – rests on the wrong assumption that reactive attitudes are voluntary in the way that actions are voluntary. The reactive attitudes "are, like belief, attitudes for which we can be asked our reasons, but which we cannot adopt at will."[44] If they were voluntary and could be adopted at will (the way one can choose at will to punish or not punish someone), they could be directly responsive to extrinsic reasons, including reasons of merit. Instead, reactive attitudes are nonvoluntary responses that reflect what matters to us by presenting the object of the reaction (such as our own act of killing someone) as having a certain (dis)value (to us). It is what matters to the person who has the self-reactive attitude that provides reasons of the right kind – constitutive reasons – for the self-reactive attitude; these reasons are considerations that bear on the question of what matters about the action, but it is important to note that this does not necessarily mean that they are *good* reasons. Someone might be wrong about what matters; they could be making a mistake when their self-reactive attitude presents an action as having certain evaluative features; in that case, we could say they have *bad* reasons of the right kind. In the next section, I will consider whether fairness *must* matter to someone in order for them to not be making a mistake about what matters, that is, in order for their reactive attitude to reflect what *rightly* matters. So far, I have simply pointed out that the question of what rightly matters determines the *fittingness* of the reactive attitude, and this is a separate question from the question of whether there are extrinsic reasons for or against having a reactive attitude of blaming someone.

The Unthinkable

What rightly matters to service members and veterans about their own terrible but unavoidable actions? I am posing these questions in order to determine whether the distress expressed by morally injured veterans fits its

[43] Pamela Hieronymi, "I'll Bet You Think this Blame is About You," in Justin Coates and Neal Tognazzini (eds.), *Oxford Studies in Agency and Responsibility Vol. 5: Themes from the Philosophy of Gary Watson* (Oxford: Oxford University Press, 2019), 60.

[44] Ibid., 66. In this way, a reactive attitude such as resentment is not *about* the one who is being resented, in the sense that it is not about whether or not they deserve the resentment; rather, it is about the one who resents, in the sense that it is about what matters to them.

object, namely whether it reflects what rightly matters to them, regardless of whatever extrinsic reasons there may be for them to not be distressed. Another way to put this is to ask what kinds of normative expectations of themselves they may rightly hold themselves responsible for meeting. Putting it this way may help us determine whether what people normatively expect of themselves can justifiably differ from what other people normatively expect of them.

What I want to suggest is that in some situations, it is fitting to have a self-reactive attitude that presents our treatment of others – for instance, killing someone – as *unthinkable* (in a sense to be elaborated). Morally injured veterans who hold themselves responsible for unavoidable acts such as killing may be expressing exactly this: that taking a human life remains unthinkable, even when it was the best thing they could have done. They may normatively expect themselves to avoid doing the unthinkable, even when the unthinkable is unavoidable, though this is not something that other people may normatively expect of them and thus hold them responsible for.

Strawson characterizes what matters to us in interpersonal relationships as goodwill, or at least lack of ill will, and other theorists following Strawson have focused on empathy.[45] But how we expect normal empathy to manifest in the way people treat each other depends on the level of vulnerability of one person to another. An empathic person who holds another's life in their hands, and who does not dehumanize the other person in any way, will generally find it to be *unthinkable* to kill the other person; in more intimate interpersonal relationships (perhaps between fellow soldiers rather than between a soldier and an enemy) someone may find it unthinkable to sacrifice the other person's life by failing to save them or letting them die. Faced with a situation in which we might have to kill or leave someone to die, we may have the experience of what Harry Frankfurt

[45] Other theorists working on the topic of reactive attitudes have developed the notion of quality of will in detail and have considered what qualities of will we demand from others. For instance, Shoemaker's tripartite division of senses of responsibility – attributability, answerability, and accountability – is organized so that each sense of responsibility corresponds to something different that we care about in the quality of other people's wills: their character, their judgment, or their regard; Shoemaker, *Responsibility from the Margins*, p. 24. Different reactive attitudes are fitting responses to the three different "objects": it is fitting to feel admiration/disdain in response to someone's quality of character, approval/disapproval in response to someone's quality of judgment, and gratitude/anger in response to someone's quality of regard; Ibid., p. 26. For an agent to be an appropriate target of anger, the agent must be capable of the quality of regard that we seek in our interpersonal relationships, which Shoemaker describes as the complicated form of empathy that is necessary for the agent to avoid slighting others.

labels "volitional necessity," in which we cannot bring ourselves to will something because we find it to be unthinkable.[46]

Obviously, this response – finding it to be unthinkable to kill another person – can be overcome. Military training would be ineffective if it could not overcome the prohibition against killing. But, as the phenomenon of moral injury is making evident, many people who overcome the prohibition in the sense that they do manage to kill do not overcome the feeling that they are doing – or, if it is in retrospect, have done – something unthinkable. What so many of the morally injured veterans express is that they consider themselves to have transgressed a moral requirement, even though they "had no choice" – this *is* an expression of the unthinkability of their action, for to find it to be unthinkable to do something is different from just finding it to be wrong to do something: when something is unthinkable, we take ourselves to be morally prohibited from doing it even if it is impossible not to do it.

Just as a joke is funny whenever amusement is a fitting response to it, a colleague's promotion enviable whenever envy is a fitting response to it, and a wolf fearsome whenever fear is a fitting response to it, an action can be *unthinkable* whenever the experience of volitional necessity is a fitting response to it, namely when someone's having this response reflects part of what rightly matters to them. Unthinkability can be thought of as an evaluative feature of certain actions. The fitting response consists of taking ourselves to be nonnegotiably prohibited from performing the unthinkable action; put differently, in experiencing volitional necessity we find certain actions to be impossible for us to will. Since we experience it as impossible – *in the sense of unthinkable* – to will these actions, *not* performing them is required even if it is also impossible, in the more ordinary sense of "impossible." When we take an *unavoidable* action to be unthinkable, then when we unavoidably perform the action (as service members do in

[46] Harry Frankfurt, "The Importance of What We Care About," in *The Importance of What We Care About* (New York: Cambridge University Press, 1988), 86. This is not only true of civilians who never encounter a situation in which they might have to kill someone, but also true of combat soldiers – it is part of what training must overcome, but training may be only partially successful in this respect. Dave Grossman's *On Killing* reports in great detail on the psychological resistance to killing that soldiers experience and explains how military training can and does function to overcome this resistance; Dave Grossman, *On Killing: The Psychological Cost of Learning to Kill in War and Society* (New York: Back Bay Books, 1995). Although Grossman's work draws in part on a study by S.L.A. Marshall that has been called into question (particularly his claims about the numbers of soldiers who could not bring themselves to fire their weapons to kill) Grossman's main points seem to hold. For an interesting study of people's (civilians') "aversion to harmful action," see Fiery Cushman, et al., "Simulating Murder: The Aversion to Harmful Action," *Emotion*, 12, no. 1 (2012), 2–7.

situations, for instance, in which killing is the best thing they can do), we must take ourselves to have failed to meet the impossible requirement not to perform the action. The recognition of this failure is what is reflected in the distressed emotions of moral injury. The distress is a fitting self-reactive attitude because it correctly apprehends the evaluative features of our having done what we unavoidably did – namely the unthinkability of it – and with it, our own failure. Even if we had no control over performing the action, we have failed to meet the impossible requirement not to perform it. The anguished sense of responsibility characteristic of moral injury reflects the recognition of the failure to meet the normative expectation not to do the unthinkable, even if it is unavoidable.

The normative expectation that someone avoid the unavoidable, however, is an unfair expectation. Is this relevant? I believe that, in general, unfairness matters when we apply the normative expectation to other people by reflecting it in our interpersonal reactive attitudes, but not necessarily when we apply it to ourselves by reflecting it in our self-reactive attitudes. In short, we are permitted to have unfair expectations of ourselves. This is because what rightly matters to us – what we can hold ourselves responsible for – includes two kinds of values: values that we share with other members of our society, and solely subjective values. We may have a subjective experience of valuing that is not tempered or limited by concerns about fairness, and this may be the basis for the normative expectation that one avoid doing something unthinkable even if it is impossible to avoid. Subjective experiences of valuing might also lead one to have expectations of oneself for levels of self-sacrifice that others would regard as supererogatory and that others could not justifiably expect of one.[47]

In contrast, normative expectations may only rightly come to be shared in certain ways, and an interpersonal reactive attitude is a demand for someone else to meet either what *are* shared normative expectations or what one, by virtue of having a particular interpersonal reactive attitude, is proposing *should* be a shared normative expectation. Norms of fairness, reciprocity, or justifiability are often cited – for instance, in contractualist

[47] Bernard Williams has made this point: "Ethically outstanding or possibly heroic actions…, in being more than obligations, are not obligatory, and we cannot usually be asked to do them or be blamed for not doing them. But the agent who does such a thing may feel that he must do it, that there is no alternative for him, while at the same time recognizing that it would not be a demand on others. The thought may come in the form that it is a demand on him, but not on others, because he is different from others; but the difference will then typically turn out to consist in the fact that he is someone who has this very conviction"; Bernard Williams, *Ethics and the Limits of Philosophy* (Cambridge, MA: Harvard University Press, 1985), p. 188–89.

moral theory – as regulating a hypothetical process through which further normative expectations can be constructed; one way to construe this is to say that people may not claim a normative principle ought to be shared if others, properly motivated by the aim of finding shared principles, could reasonably reject it.[48] Despite the fact that norms of fairness may in practice be violated, or that the concept of fairness can be interpreted in conflicting ways and does not necessarily imply equality, it is still the case that many actual societies have developed norms of fairness or reciprocity that play some role in regulating the construction of further norms.[49] Our interpersonal reactive attitudes must either reflect these fairly constructed normative expectations or serve to fairly propose that some normative expectation be shared.

We may hold ourselves responsible for more than what we may hold most other members of our society responsible for, because our expectations of ourselves may be based on a subjective experience of valuing that is not tied to such a process of norm construction, and thus the self-reactive

[48] T.M. Scanlon, *What We Owe to Each Other* (Cambridge, MA: Harvard University Press, 1988). This does not mean that actual people do not try to hold others responsible in ways that reflect unfair or unjustifiable normative expectations or that actual people do not (perhaps in a self-deceived way) put forward claims that appear to be fair when they are not. Shared normative expectations may very well be unfair (often because they apply differently to members of different social groups); and people may very well have interpersonal reactive attitudes that reflect what they believe should be shared normative expectations, even though these expectations would be unfair. The common phenomenon of "blaming the victim" is based on unfair normative expectations; this is evident, for example, when a woman is held responsible for being raped because she was walking alone at night or dressed in a way that the rapist found to be provocative, or a black man is held responsible for being subject to excessive force during an arrest because he acted in a manner that caused a police officer to feel threatened.

[49] One can argue (on the basis of a claim about the function of morality, given certain assumptions about human nature and about what human cooperation requires) that any moral system that does not have some norm of fairness or reciprocity, and some recognition of the need for justification of shared normative expectations, is inadequate. David Wong, in developing a constructivist position that he calls *pluralistic relativism*, argues that there are some universal constraints on any adequate morality, including that any adequate morality must have some norm of reciprocity and some requirement that norms be justifiable to those who are to be governed by them; David Wong, *Natural Moralities: A Defense of Pluralistic Relativism* (New York: Oxford University Press, 2006). He claims that "some form of reciprocity is a norm for all cultures we know, where reciprocity is conceived as a fitting and proportional return of good for good" (ibid., p. 47), noting that, "while reciprocity in a very general sense is the norm for human cultures, the specific form it takes varies a great deal across cultures"; ibid., p. 50. Regarding the constraint of justifiability, he points out that "as a system for promoting cooperation, morality works through a large degree of voluntary acceptance of its norms... If conformance to its norms...depended solely on the threat of force or coercion, the costs would detract greatly from the benefits of social cooperation itself... voluntary acceptance of moral norms came to be seen as based on their justifiability to those governed by them. Hence another constraint on moralities is that justification for following the norms and reasons of an adequate morality cannot crucially depend on falsehoods"; ibid., p. 59.

attitudes characteristic of moral injury can be fitting. I may not justifiably impose on others expectations based on what matters only to me, that is, based on values I could not expect them to accept, but I may hold myself to these expectations, for I *have* in fact accepted them. When, through self-reactive attitudes, I hold myself responsible specifically in relation to my subjective values, the fact that I hold *only* myself responsible means that there is no chance of holding someone responsible for values that they neither share nor could be reasonably expected to share. For shared values, the question to be settled is not what matters to *me*, but what matters to *us*, and if determining what matters to us entails knowing what we can reasonably expect each other to accept, shared normative expectations will be limited in ways that my expectations of myself are not.[50] While I can certainly hold others responsible for not killing when killing is an action that they *can* avoid, I cannot reasonably expect others to avoid the unavoidable or to emerge from a moral dilemma without committing an action that would otherwise be prohibited. But through the subjective experience of volitional necessity, I may very well require these sorts of impossible things of myself.

Conclusion

How do we answer practical questions about how to respond to the problem of moral injury and its associated distress, if, as I have argued, the distress is fitting, and yet there are extrinsic reasons that recommend against it? One part of the answer is obvious: we should undoubtedly work to reduce the *causes* of moral injury, whenever they can be reduced. Moral injuries experienced by perpetrators of *avoidable* wrongdoings can be prevented to the extent that the perpetrators can be prevented from committing the wrongdoings. Even regarding wrongdoings that were unavoidable *given the situation someone was in*, it may be that their being in that situation in the first place was completely avoidable: bad political decisions (including the decision to go to war), inadequate efforts at diplomacy, and poor military leadership can all result in service members being put in situations in which they have no better options than to commit moral wrongdoings. Progress in democratic politics, better diplomacy, and improvements in

[50] Scanlon's contractualism leaves him room to accept what I am claiming here, because he acknowledges that the contractualist theory applies only to one circumscribed area of morality – the realm of "what we owe to each other" – and that morality more broadly understood admits a greater plurality of values; Scanlon, *What We Owe to Each Other*.

military leadership could prevent some moral injuries.[51] Nevertheless, it is a sad truth that not all of the causes of moral injury can be eliminated, because moral failure is not always avoidable: there will always be moral luck, epistemic limitations, and conflicts between incommensurable values, that lead in different ways to unavoidable failures, as well as situations that force people to choose between wrongdoing and self-sacrifice. I am not a pacifist and do not believe either that war can always be avoided or that war can be conducted without warriors having to commit acts that it is fitting to treat as unthinkable.

A separate question concerns how others should respond to a veteran who expresses distress once a moral injury has taken place and the veteran is suffering from it. In the case of veterans whose moral injury stems from a wrongdoing that they could have and should have avoided, both self-blame and blame from others are fitting and serve to hold the person responsible for meeting fair normative expectations. Here the therapeutic approach called Adaptive Disclosure makes sense conceptually.[52] This approach involves the therapist affirming the appropriateness of the veteran holding themself responsible for their wrongdoing: it "allows for accurate and legitimate self-blame when warranted but also promotes the possibilities of self-forgiveness, compassion, and moral reparation",[53] and "encourages a recommitment to pre-event personal ethical and moral standards."[54]

Responding well to a veteran whose moral injury stems from their experience of *unavoidable* moral failure, on the other hand, depends on accepting the fittingness of the veteran's attitude of holding themself responsible for the wrongdoing, while meanwhile recognizing that there are extrinsic reasons for diminishing the distress. The challenge is to reduce the suffering *without* engaging in any denial of the fittingness of the self-reactive

[51] Preventing moral injury by improving military leadership is one of the main aims of Jonathan Shay's work; Shay, *Achilles in Vietnam*; *Odysseus in America*; and "Moral Injury." While he employs a definition of "moral injury" that is focused on how one can be injured by being betrayed by people who are in a position of authority, he also recognizes that such betrayals, or failures of leadership more generally, can put service members in positions in which they themselves must commit wrongdoings.

[52] See Litz. et al., "Moral Injury"; Shira Maguen and Brett Litz, "Moral Injury in Veterans of War," *PTSD Research Quarterly*, National Center for Posttraumatic Stress Disorder, 23, no. 1 (2012), 1–3; and Gray, Nash and Litz "When Self-Blame is Rational and Appropriate." While I endorse the conceptual assumptions behind this approach, I am not in a position to evaluate the approach empirically or to say anything about its therapeutic effectiveness.

[53] Gray, Nash and Litz "When Self-Blame is Rational and Appropriate," 383.

[54] Ibid., 386.

attitude, and thus to leave room for them to hold themself responsible and to express that they have failed morally.[55] In insisting that some of the actions that it may be beyond our control to avoid are truly unthinkable, morally injured veterans are actually offering anyone who will listen to them an important but tragic truth that we should treasure even if we would prefer the fantasy that moral goodness is always possible.

[55] Bryan Doerries' "Theater of War," I believe, may achieve exactly this kind of response to morally injured service members and veterans. See https://theaterofwar.com/.

CHAPTER 3

Stoic Grit, Moral Injury, and Resilience

Nancy Sherman

Case of Accidental Killing

Layne McDowell was meant for a cockpit. He had wanted to fly since junior high, and the United States Naval Academy, unlike the Air Force, took a gamble on a guy that had just had a knee injury. He graduated in 1995 with a 3.84 Grade Point Average and soon discovered he also had the physiology to fly. He was a gravitational force monster, able to withstand 9-g over time in the "spin and puke" centrifuge. He could physically endure and had made peace with willful self-defensive killing of enemy combatants in what he took to be just war ways. But one event early in his career unhinged his sense of moral calm. It was a midday strike on a radio relay site in Northern Kosovo in May 1999. Intelligence imagery was grainy. In order not to alert Serbian forces, he had to go south of the target and make a quick turn back. Aircrews now had less time to locate and verify the target. Serbian Air Defense opened fire and that took McDowell's attention away from the targeting screen.

"I felt good about the release. Then clouds obscured the target until about 13 seconds to impact. At that time I began having doubts about the target. It didn't look right, but in those 13 seconds, I didn't say anything, and we took out what we were targeting with 2 GBU [guided bomb unit]-12's."

Dread started to mount. Back on the carrier, McDowell looked at the strike footage on a big screen. The bomb had struck not the target but a carport next to a house. McDowell saw signs of civilian occupation and, unmistakably, four bikes, two of which were child-sized.

There were never any legal proceedings or Navy follow-up to determine if, who, and how many civilians and children may have been killed in the strike. But he carried the moral burden in a repetitive intrusive dream in which he did his own after-incident investigation. The dream replayed again before he deployed to Iraq in 2005. The building he bombed was somehow still standing but there was thick dust everywhere, insulation and wires dangling, boards littered all over the ground. The smoke was thick and it was hard to make out who, at all, was in the structure. He aches to turn back the clock, to be given time to steer the bomb to an empty field. But he can't. In the structure, he definitely saw a small boy huddled in the corner, coated in dust, severely injured but still breathing. He knew the face. It was McDowell's own son, Landon. "He lifted the boy to his chest, tightly for a hug, cupping his hand behind the child's little head, to hold it. The back of his skull was gone."

A Tension: Moral Injury vs. Stoic Resilience

The case I have retold is drawn from American journalist and Marine C.J. Chivers' *The Fighters.*[1] It is not of collateral killing of noncombatants, but of accidental killing. Unlike some collateral killings that may be justified as necessary militarily, and which are excused as part of eliminating a serious threat where there are no less harmful means for taking out the target or risks that can be transferred to fighters without thereby undermining a mission, accidents like these, all too numerous in war, are never justified as necessary or eliminative killings.[2] There is no military good to be achieved: the noncombatant killing is not part of a proportionality calculation. The act is transgressive of someone not liable to be killed. Even if an individual's acts are legally or morally excusable – as here, due to poor intelligence, sudden blinding by enemy air fire, unpredictable shifts in flight patterns or cloud cover ("fog of war" excusing facts) – innocents have nonetheless been wronged. In McDowell's case, as proximate causal agent, he carries that moral burden. In his accounting, he had thirteen seconds to readjust the targeting. Doubt crept in during those seconds. He was the aviator who dropped the ordnance. His punishment comes in the recurrent nightmare, in his wishing to turn back the clock and steer the bomb differently, in growing lukewarm over time about flight missions, in his yearning to know who was killed so he and the Navy can make atonement through compensation.

This is a striking case of military moral injury. Yet what is puzzling in this case is that the transgression that caused so much moral distress may

[1] C. J. Chivers, *The Fighters: Americans in Combat in Afghanistan and Iraq* (New York: Simon & Schuster, 2018), pp. 6–24, 119–21.

[2] Seth Lazar, *Sparing Civilians* (Oxford: Oxford University Press, 2015), p. 60. I am grateful to Michael Gross for conversation about moral injury and puzzling cases. For discussion of transfer of risk, see Tony Pfaff, *Resolving Ethical Challenges in an Era of Persistent Conflict* (Scotts Valley, CA: CreateSpace Independent Publishing Platform, 2011).

not have run afoul of the moral and legal codes of war. The agent can't exculpate himself, though the doctrine of war can. We could say this is just the grandiosity of moral agency at work, especially of warfighters who are raised on perfection and zero-tolerance for mistakes, whether in keeping a rifle clean or in hitting a target with expensive precision munitions. Many hold themselves "strictly liable" even when they ought not. But shunting all or most military moral injury to psychological overreach misses the greater moral landscape. Even if we can't always count on sensitivities to accurately track what we morally should and shouldn't do, still, for those who aren't morally callous, the array of emotions experienced with moral injury – guilt, shame, moral indignation, resentment, betrayal, a yearning to atone – speak to aspiration.[3] Moral anxiety is inseparable from the desire to be good and to do better, even in the most cramped circumstances and in the face of horrible tragic luck. If those who go to war didn't feel anguish at the detritus of war, whether incurred by act or omission, suffered or witnessed, we'd wonder about their humanity.

This brings me to my central concern. There is a glaring tension in military culture between a recognition of moral injury and repair as a healthy expression of the moral conscience of a soldier, and a strong attraction to a Stoic ideal of virtue and resilience that distances itself from the disquiet of emotions – such as guilt and shame, resentment and indignation, mercy and forgiveness – that are constitutive and expressive of moral injury and repair. This forces the question: Is moral injury compatible with a popularized Stoic view of moral strength as tough grit? Does that view actually do justice to Stoic doctrine? And if it doesn't, what can we learn from a more nuanced account of Stoicism?

My concern here is not just academic or textual. It has to do with education and what we read and take from that reading. If Roman Stoics, such as Epictetus, Marcus Aurelius, and Seneca, and redactors, such as Cicero, are sources of inspiration for training military virtue and resilience, are we reaping the right lessons when we read their writings? Are there other Roman Stoics we should also be reading and teaching? With a broader and deeper grasp of ancient Stoicism, can we construct a healthy modern Stoicism that both recognizes moral injury and offers lessons for healing?[4]

In this chapter, I argue "yes," and turn to Seneca to develop my answer. At military academies, Seneca is not the Stoic favorite that Epictetus or

[3] See Nancy Sherman, *Afterwar: Healing the Moral Wounds of Our Soldiers* (New York: Oxford University Press, 2015).

[4] I develop the answer more fully in Nancy Sherman, *Stoic Wisdom: Ancient Lessons for Modern Resilience* (New York: Oxford University Press, 2021).

Marcus Aurelius is. But that's a pity. For not only is he a dazzling writer, he often shares how hard it is to live virtuously in an ecosystem that constantly tests his moral resolve. His struggle speaks to many in the military world where chains of command and cumbersome bureaucracy can often compromise wise moral choice.

The plan of the chapter is this: First I limn the broad lines of a philosophical approach to military moral injury against the backdrop of clinical research. Second, I consider whether there is normative space in Stoic writings for the emotional disquiet of moral injury, especially in the context of moral aspiration. Here, I draw from Cicero's redactions of the Stoics. Finally, I ask: How would the Stoics counsel us to calm the anguish of moral self-blame, especially when that self-condemnation is crippling and may not accurately track real wrongdoings in war? To develop my answer, I turn to Seneca's reflections on mercy and benevolence in his essay *On Mercy*, and in his plays *Hercules Rages* and the *Trojan Woman*. Mercy, Seneca insists, makes good on the gentler side of Stoicism. I argue that learning from the mercy others show us or that we expect they would show us, and that we, in turn, would show them, is one way that soldiers can begin to show mercy toward themselves.

Moral Injury and Reactive Attitudes

The leading research and clinical mental health professionals working on war-related moral injury define it as "a syndrome of shame, self-handicapping, anger, and demoralization that occurs when deeply held beliefs and expectations about moral and ethical conduct are transgressed." They argue that the transgressions can arise from the point of view of the agent, from the behavior of others, or by bearing witness.[5] A critical piece of the research insists that moral injury is distinct from fear-based trauma, or traditionally understood posttraumatic stress disorder (PTSD), in that the potentially injurious event is not perceived as an overwhelming life threat, but as a severe moral conflict or challenge.[6] The experiences have

[5] Brett T. Litz et al., *Adaptive Disclosure: A New Treatment for Military Trauma, Loss, and Moral Injury* (New York and London: Guilford Press, 2016), p. 21. See also Brett Litz et al., "Moral injury and Moral Repair in War Veterans: A Preliminary Model and Intervention Strategy," *Clinical Psychology Review*, 29, no. 8 (2009), 695–706; and Shira Maguen et al., "Killing and Latent Classes of PTSD Symptoms in Iraq and Afghanistan Veterans," *Journal of Affective Disorders*, 145, no. 3 (2012), 344–48.

[6] I have used the term "posttraumatic stress disorder" and "PTSD" for the sake of consistency because the literature to which I am referring uses that term. That said, I am sympathetic to many in the military and beyond who have argued that the "D" for "disorder" can unduly stigmatize psychological trauma.

to do with breaches of morality, not breaches of safety. Moral injury is also distinct from complicated grief reactions in experiencing loss. Still, the symptoms of moral injury can overlap with those of PTSD, such as re-experiencing injurious experiences (as in flashbacks), strategic avoidance of those experiences, and emotional numbing.[7] The diagnostic criteria may also overlap, as in the most recent edition of the *Diagnostic and Statistical Manual of Mental Disorders* (V), which in addition to fear lists the emotions of anger, guilt, and shame as diagnostic criteria for PTSD.

My own work is not as a clinician (though I have psychoanalytic research training and have at times worked closely with clinicians). I have long argued that it is wrong to think of *all* military moral injury as pathological or maladaptive. This is not at all to minimize the critical work of behavioral or mental health therapy. If anything, far more needs to be done to destigmatize mental health interventions in the military as a whole and beyond it, nationwide and worldwide. Within the military, it is crucial that leaders own their own mental health challenges and lead troops by example, as they do in other ways.

Still, a mono-lens of diagnostic criteria drawn from psychopathology obscures the fact that some mild moral injury can be salutary for both the individual service member and for the profession of arms as a whole. Moral injury can signal a morally conscientious soldier working out consciously or subconsciously the contours of moral accountability in war, even if blame is wrongly attributed at times. Equally, moral injury, however mild or severe (and whether triggered by actual transgressions or simply moral conscientiousness) may never get fully processed within or outside therapy before it is time for a soldier to prepare to deploy again and ramp up to operational tempo mode. But even unprocessed moral disquiet may be an important first step in an individual reaffirming their sense of personal accountability in war and in strengthening the sensitivities that are a bulwark against heinous wartime transgressions.

Here, think of a fighter feeling pangs of guilt in the wake of permissible collateral killing during a strike against a high-value target (say in the case of the strike that killed terrorist mastermind Abu Musab al-Zarqawi, where it's likely it was known in advance that there were family members with him). Or a soldier feeling sullied in shame in the wake of consummating a high-stakes political deal with a shady but allied warlord. Or a soldier weighed down by both a sense of command betrayal and personal shame in being ordered to issue a paltry consolation payment to a civilian family

[7] Litz, "Moral Injury and Moral Repair."

caught in the crosshairs of a return strike. In these cases the emotions are apt: guilt at an undeserved civilian killing, shame in accepting the terms of a corrupt power broker, and moral disappointment in those who are meant to be exemplars in conducting a population-centric war and shame in carrying out their shoddy work. The emotions look backward but also forward and are wide-lensed. To feel their bite is to hold onto aspirations for what goodness might look like in less constrained circumstances. It is to hold onto a more capacious conception of moral character and goodness than the persona of a just combatant often easily accommodates.

In all this, I am framing morally injurious emotions in terms of Strawsonian "reactive attitudes."[8] The core idea is that we have a basic concern and demand to be treated with "regard and goodwill" and that this expectation displays itself in certain positive and negative moral sentiments that help to explain our practices of attributing responsibility.[9] Strawson focuses our attention first on particular personal attitudes of gratitude and resentment that we feel and typically express in reaction to actions of others that show intentions of good or ill will directed toward us. From these personal reactive attitudes, Strawson expands the range of emotions to include vicarious reactive attitudes felt and expressed on behalf of others, as in the case of "indignation," Strawson's stipulative replacement term for resentment. Finally, Strawson turns the direction of reactive attitudes toward self and those reflexive emotions felt and expressed when we are the agent of good or ill will. So, we feel guilt or shame in actions that manifest ill will toward ourselves or others, or on the positive side, renewed trust, forgiveness, or empathic concern when we take actions to atone and repair.

I have developed a conception of moral injury as reactive attitude in *Afterwar*.[10] I draw on that work here as background: At the heart of suffering moral injury are "responsibility practices" manifest in distinctive emotions. To lead with moral emotions and the responsibility practices that they help explain, rather than psychiatric diagnoses and disorders, is to shine a bright light on the moral dimensions of military psychological trauma and the critical need for mental health interventions that squarely address those moral dimensions. In all this it is critical to note,

[8] P. F. Strawson, "Freedom and Resentment," *Proceedings of the British Academy*, 48 (1962), 1–25. For an insightful reconstruction of Strawson's views and those of interpreters, such as Darwall and Wallace, see John Deigh, *From Psychology to Morality: Essays in Ethical Naturalism* (New York: Oxford University Press, 2018), pp. 101–28.

[9] Gary Watson, "Peter Strawson on Responsibility and Sociality," in David Shoemaker and Neal Tognazzini (eds.), *Oxford Studies in Agency and Responsibility, Vol II: 'Freedom and Resentment' at 50* (Oxford: Oxford University Press, 2014), 17.

[10] Sherman, *Afterwar*.

that moral injury is suffered in war no less than outside war; transgressions, committed and suffered, as well as acts that fall short of endorsed ends of good character, can give rise to moral anguish and a need for repair. What we learn from the military about the importance of recognizing and addressing moral injury has important ramifications for all of us.

Stoic Emotions and the Moral Progressor

The Stoics don't talk about moral injury *per se*, but they do talk about moral distress. And they teach that such distress has no place in the personality profile of the ideal moral person. For with Socrates, they hold that the only real harm is to become unjust, and that is not a possibility for the truly virtuous person. A truly good person, teaches Socrates in the *Apology*, cannot be harmed in life or death. In other words, such a person cannot be morally injured. For the Stoics, this idealized person is the sage.

The Stoics fill out the details. Bad fortune, loss of loved ones, and physical or mental injury, even being wronged by others, don't affect genuine happiness or wellbeing (*eudaimonia*). We don't wish these things to happen, but all too often, they lie beyond our control. Our own moral wrongdoing, by contrast, is within our control. Or at least, the Stoics have us focus on pure cases where our hands aren't forced. They draw a bright-line rule for a sage's behavior: A sage can do no wrong by definition. And so, there is no room for moral anguish or angst. The absence of this emotion makes for a glaring hole in the Stoic classification of "good" or healthy correlates of the four basic emotions. Ordinary emotions (pleasure, desire, distress, and fear) and their subspecies are, by and large, preferences or dispreferences for external goods (what the Stoics call "indifferents"). Their danger is that they can easily give way to excess in restless yearning or panicky aversion. Good emotions, cultivated by the sage, are equable correlates, directed at the pursuit of virtue and the avoidance of vice. Thus, a sage no longer experiences full-throated ordinary desire for external goods, but a rational wish to pursue virtue. Similarly, full-throated fear yields to rational caution in avoiding vice, and pleasure to serene joy in the fulfillment of virtue. But significantly, there is no analog for distress. For a sage does nothing that would bring on feelings of genuine moral distress or anguish.[11]

[11] A. A. Long, D. N. Sedley, & David Sedley, *The Hellenistic Philosophers: Volume 1, Translations of the Principal Sources with Philosophical Commentary* (New York: Cambridge University Press, 1987), 65E–F, p. 412.

But what if you are not a sage? The sage often rises only as the phoenix, about every 500 years. Most of us will not rise to that exalted, divine height. And so, what if you are, like Seneca says he always is, just a moral progressor, aspiring to become better but subject to error, misevaluations of what is really worthy, caught in struggles with those in power who compromise your moral autonomy and self-rule? To bring the case closer to home, what if you are a military fighter, like Layne McDowell, part of a chain of command, dependent upon a vast network for training and operation for intelligence collection and analysis, and much more? The Roman Stoics understand well that agency is framed and nested within systems, and not all of them healthy or benign. That is one attraction of the Stoic retreat that makes goodness a matter of inner virtue. That's Epictetus's tact, an enslaved Roman who finds real freedom, both when enslaved and emancipated, in inner mastery. But Seneca is a public servant. He wrestles with the demands of virtue in the arena of politics and hierarchical power. And this makes him a compelling figure for thinking about the military actor whose agency is framed by a chain of command and complex, often byzantine bureaucracy.

In this regard, it's worth recalling briefly who Seneca is and whom he served, in order to appreciate why moral aspiration and striving frame many of his moral writings. Seneca the younger (3 BCE – 65 CD) (the Seneca most of us know), was a public philosopher, who, though he may not have known the fog of war, knew well the muddy waters of politics. He swam in those currents as Nero's tutor, political adviser, and speechwriter. Nero, the 16-year-old boy emperor, didn't necessarily want to repeat the reign of Caligula, but he didn't hold back when the dynastic succession was threatened. Seneca thought he could help him and restrain his more wanton ways. But ultimately, he wasn't so successful. A few historical details should suffice: Seneca came to the court through the interventions of young Nero's mother, Agrippina, who sought him out as a tutor with the most famed reputation in the Empire for rhetoric and public speaking. She became his patron, bringing him out of exile in Corsica (41CE) where her husband Claudius had banished him for eight years for his alleged adultery with his niece, Julia Livilla. Getting Nero closer to the throne through the powerful pen of Seneca was part of her motive. Seneca's motives then became keeping the boy in power, even though it soon involved keeping the mother, who had fallen out of favor with her son, out. There was no shortage of intrigue in this palace where Seneca lurks in the background. There was the poisoning of Claudius's biological son, Britannicus, when he became of age to take the throne, a plot of which Seneca, as a palace insider, was likely aware. Though Agrippina was his benefactor in getting him to the palace, he showed her limited gratitude, defending Nero's murder of her in a speech that records suggest didn't go down well with all.[12] What goes around comes around, and while Nero's anger may have been restrained on occasion, it was by no means ever managed. Seneca must have been well aware of the threat to his own life, given his attempts to retire from public life in his later years and his general preoccupation in his writings in those years (notably, in the *Letters*) with issues of mortality and the transience of power. In 65 CE, Nero ordered Seneca's suicide on charges that he was involved in the Pisonian plot to have the emperor assassinated.

This biographical sketch reminds us that when Seneca writes about anger or clemency or moral constancy, or rails against the evils of materialism and glory, he did not write as a moral or political naïf. He knew the pull of wealth and power and the perils of trying to escape it under the eye of a watchful and vengeful tyrant. He took up his pen and well-trained Stoic stance, in part, to calm his own fears about political power (and those likely of many in his audience) and aspire to something purer. As he often said, he wrote from the perspective never of a sage but from that of the moral doctor who was at once a patient, in need of Stoic medicine and healing. He rose from a modest equestrian background in the provinces, and never, like some other Stoics in Rome, stayed clear of Nero's inner circle and, once in it and wanting out, opted for his own suicide rather than being forced to it.[13] In short, he was a pragmatist who had been in the political trenches, suffered its glory and infamy, and especially, in later years, yearned for personal moral change. That, at least, is a way to begin to read some of his moral essays, letters, and tragedies.

For many of us (though maybe not all in our contemporary political scene) moral compromises may not rise to the level of imperial court intrigue, with execution, poisoning, banishment, imprisonment, and enforced suicide looming in the background or foreground. But the basic condition of not being sin-free and yet aspiring to become better is, in part, what has appealed to readers of Seneca throughout the ages, in the Hellenistic world and the Judeo- Christian period that followed. And it also is part of the implicit appeal of Stoicism for the military. For their

[12] For this snapshot biography of Seneca, I draw from Emily Wilson, *The Greatest Empire: A Life of Seneca* (New York: Oxford University Press, 2014).

[13] In this regard, Seneca does not write from the point of view of an Epictetus, formerly enslaved and impoverished all his life, or that of Musonius Rufus, Epictetus's teacher.

culture is one not just of unbridled can-do-ism, but of constraint and chains of authority that can squeeze autonomy and force choices that leave moral detritus in their wake. Moral injury, aspiration, and repair are part of that culture, even if not always openly acknowledged.

But where exactly do we find this thread of moral aspiration in Stoic writings? We first have to go back to Plato. A famous redactor of Stoic texts, namely Cicero, recalls this scene from the *Symposium.*

The Tears of Alcibiades

At the conclusion of Plato's *Symposium*, at a banquet in honor of the god Eros, Alcibiades, the morally flawed and disastrous military leader who betrayed Athens to the Spartans, bursts into the drinking party and addresses his love encomium directly to Socrates, his beloved moral tutor. Socrates, he confesses, is the only one who can really hold up a mirror to his errant ways and bring on the tears of shame. The anguish is, at times, excruciating, especially in Socrates' presence. For at those moments, says Alcibiades, "He always traps me, you see, and he makes me admit that my political career is a waste of time, while all that matters is just what I most neglect: my personal shortcomings which cry out for the closest attention." "Socrates is the only man in the world," he says baring his soul, "who has made me feel shame." "Ah –," he says knowing his audience, "you didn't think I had it in me, did you? Yes, he makes me feel ashamed."[14]

Alcibiades is a tormented soul. He's not a typical weak-willed or *akratic* person, as Aristotle understood the term – someone who knows clearly what he should do but doesn't do it. Alcibiades doesn't have firm moral principles that get trumped on occasion, through temptation or self-deception or the like. He's far more ambivalent. He occasionally dips his toes into the waters of virtue, especially when Socrates is looking on and casting judgment. But, as he confesses, he hasn't really dedicated himself to the hard work of building new habits and weaning himself from old ones.[15] Military glory and political fame still hold sway for him. He needs external prods and sanctions to keep him on course. In those moments, with Socrates *in vivo*, or vividly in mind, he feels the deep shame of his "old ways" and of having too often caved into desires "to please the crowd." And that is a spur to do better.

[14] Plato, *Symposium*, trans. Alexander Nehamas and Paul Woodruff (Indianapolis, IN: Hackett Publishing, 1989), pp. 216a–b. See an insightful study, see Margaret Graver, *Stoicism and Emotion* (Chicago: University of Chicago Press, 2007), pp. 191–211.

[15] See Agnes Callard, *Aspiration* (New York: Oxford University Press, 2018), pp. 14–31.

The "tears of Alcibiades" becomes a challenge for Stoic thought: how to understand moral distress as a part of moral improvement? Cicero poses the challenge. Not himself a Stoic but committed to preserving Hellenistic philosophy in the Roman world, he is often attracted to Stoic ways, but also critical. In the *Tusculan Disputations*, Cicero insists that Cleanthes, the second of the three Greek patriarchs of the Stoic school, doesn't take the problem seriously enough: "It seems to me that Cleanthes does not take sufficiently into account the possibility that a person might be distressed over the very thing which Cleanthes himself counts as the worst of evils." And then Cicero reminds his readers of the *Symposium* passage: "For we are told that Socrates once persuaded Alcibiades he was unworthy to be called human, and was no better than a manual laborer despite his noble birth. Alcibiades then became very upset, begging Socrates with tears to take away his shameful character and give him a virtuous one." Cicero presses the Stoics to make sense of Alcibiades's tears: "What are we to say about this, Cleanthes? Surely you would not claim that the circumstances which occasioned Alcibiades' distress was not really a bad thing?"[16]

Cicero hammers home the point later in the same essay: "Suppose a person is upset about his own lack of virtue – his lack of courage, say, or of responsibility or integrity. The cause of his anxiety is indeed an evil!" It is an "impulse toward virtue itself," he says.[17] Cicero admits that it can be an "all too vigorous impulse" that can lay us low. His therapeutic counsel is not to dismiss the cause of distress, but to control the outer expression. We should try to manage the tears and lack of decorum. But if the core cause and object of distress is our own wrongdoing, then we should seize the moment as an occasion for moral aspiration. Take a first step and harness the impulse toward virtue.

With this as background, let's return to military moral injury. Lane McDowell holds himself morally culpable, even if his indictment distorts or exaggerates his own responsibility for what happened. He is not alone. So, too, do so many service members that I have interviewed and written about over the years.[18] For some, the burden they bear is that they have come home from war when their battle buddies did not. They hold

[16] Cicero, *Cicero on the Emotions: Tusculan Disputations 3 and 4*, M. Graver (ed.), (Chicago: University of Chicago Press, 2002), 3.77, pp. 34–35.

[17] Ibid., 34.61–2, pp. 61–2.

[18] For more on my conversation with service members on their experiences at war and experiences of moral injury, see Nancy Sherman, *Stoic Warriors: The Ancient Philosophy Behind the Military Mind* (New York: Oxford University Press, 2005); *The Untold War: Inside the Hearts, Minds,* and Souls *of Our Soldiers* (New York: W.W. Norton, 2010); *Afterwar*; and *Stoic Wisdom.*

themselves morally responsible – for being on leave the day when IED blasts ripped through the Army vehicle of a best friend, for having squatted rather than stood on the roof the second the insurgents took aim, for having given permission to a squad mate to get out of the Humvee to relieve himself in a spot that ended up being booby-trapped with mines. Survivors' guilt, accident guilt, holding oneself morally responsible for events for which one may not even be causally responsible – this is how servicemembers carry the burden of care for each other. The guilt may be overall fitting of good character and care. You don't just feel grief: you feel you could have done something differently. Imputed agency steps in to fill a void. But the self-blame can come with a sense of failed agency that is at once too harsh and unremitting. The right therapy in these cases may involve correcting distortions and redrawing the bounds of agency. Letting go requires understanding the limits of control. But accepting those limits requires self-compassion, and that may not be so easily available.

That compassion may have to first come from others. To see this, consider a checkpoint incident in Iraq. A car with two military-aged men and a child fails to stop at the first round in a series of checkpoints close to a large, heavily populated U.S. military installation and arsenal. The car pushes through two additional checkpoints that come in quick succession without showing any signs of slowing down. The man in the front passenger seat reaches for something under the seat as the car progresses through the checkpoints. As the passenger lifts his head and straightens his torso, he seems to be cradling an explosive device. The sentry shoots just after the third checkpoint, killing everyone in the car, and seconds before what, in fact, turns out to be a bomb, detonates. As the Army sentry considered his options, he was aware that his shots might kill the child, but also aware that in giving multiple warnings to the driver and in restraining fire until the third checkpoint, he was taking considerable risk onto himself in order to minimize risk to the child and other potential innocents in the car while still securing the base.[19]

This is a hypothetical case, but not unlike those I have been told. Even if the child is being used as an involuntary shield, the soldier still feels horrific guilt at doing what he regards as unimaginable – killing an innocent child. And yet, according to the obligations of war, his action is unavoidable.

[19] Following Michael Walzer on permissible collateral killings: "Double effect is defensible ... only when the two outcomes are the product of a *double intention:* first, that the 'good' be achieved; second, that the foreseeable evil be reduced as far as possible." See Michael Walzer, *Just and Unjust Wars: A Moral Argument with Historical Illustrations* (New York: Basic Books, 1977), pp. 155–56.

His own self-sacrifice, as a way of avoiding the killing, would only lead to the deaths of many more soldiers on the base and abort the staffing and materiel for future missions.

Again, the soldier's guilt seems apt, at least in the sense of expressing a potent mix of felt remorse, sadness, and, critically, acute awareness that he was the proximate cause of the killing, however justified by rules of engagement. He doesn't feel regret, in that he feels culpable. The guilt is apt in that it captures the soldier's humanity and sense of conscientious accountability.[20] Still, it can easily become too harsh and unrelenting. One way a soldier might come to see this is by taking up an interpersonal perspective. A soldier likely wouldn't reproach another in a comparable checkpoint incident. He'd excuse him, or refrain from blame, appreciating what was an unavoidable and constrained action in the circumstances. And similarly, another might be a benevolent counselor to him, helping him see what he can't himself see or feel.

Granted, we don't easily give ourselves the same "pass" that we urge on others or others urge on us. Some of that seems right. In war, of all places, we want moral conscience to be alive, and to not numb inner sanctions, even if the meting out of them in self-judgment is sometimes just too robust. And so, the pressing practical and therapeutic question becomes: how do we cultivate ways for servicemembers to lessen and relieve guilt, so that the burden they bear is fairer?[21]

Seneca is insightful here. He counsels that learning to see from the viewpoint of others can be a way to cultivate self-empathy. This emerges in a stunning way in his play *Hercules Rages*.

Briefly, Seneca portrays the tragedy as centered on Hercules's addiction to superhero action, nursed by a madness sent by the goddess Juno. Juno is horribly jealous of Hercules, the most glorified of her husband Jupiter's illegitimate sons. Having forced twelve labors upon him, she has one final act of wild revenge. Just as he is poised to pierce through the barrier of the underworld and return home, she makes him go mad and in his crazed

[20] Bernard Williams, in laying the foundation for the moral luck literature, famously calls this kind of sentiment, "agent regret." However philosophically correct as a term of art, "agent regret" fails to capture the phenomenology, as I argue in *Untold War*. In this regard, I use "apt" to underscore the *reasonableness*, if not warrant of the guilt feelings. For related discussion on the appropriateness of emotions, see Justin D'Arms and Daniel Jacobson, "The Moralistic Fallacy: On the 'Appropriateness' of Emotions," *Philosophy and Phenomenological Research*, 61, no. 1 (2000), 65–90.

[21] Lisa Tessman's chapter in this volume attempts to reconcile the occurrence of fitting but unfair reactive attitudes in cases of moral injury. My approach here is to find in the unfairness resources servicemembers can use to recover from their injuries, no matter how fitting their feelings of guilt may be. For my own analysis of servicemembers' punishing self-blame, see Sherman, *Afterwar* and *Untold War*.

state, he slays his family. When he wakes to his hideous deed, suicidal rage is all he knows. There is no way forward. One more final, violent act will bring an end to his misery. His father enters as a benevolent counselor. He won't judge the deed: "Forgive yourself for just this one bad act."[22] Show compassion and empathy toward yourself. I am here for you to mirror mine. His beloved friend Theseus also begs him to find some path, however narrow, toward self- kindness: "Your father's prayers ought to work, but let me also try to move you with my tears." "Burst through your troubles, with your usual energy." "Use your heroic courage" to not stay angry at yourself.[23] In essence, father and friend are imploring a superhero, bent on superaction, and now final destruction, to turn his strength and courage toward mental healing. This is a profound lesson for the members of a warrior caste, who in our times have seen far too much suicide among those with whom they have honorably served.

More generally, Seneca is willing to show us raw moral distress at horrific deeds, unwittingly committed. And he is implicitly arguing that even if we motivate morality by moral shame and distress, we need others to mitigate self-blame when those condemnations are unyielding or unfair. The benevolence and goodwill of others, their forgiveness, and at times mercy, are required to nurture the self-empathy so critical for posttraumatic growth and resilience. And so, at least in this reading of *Hercules Rages*, Stoic strength and resilience are cast not as macho, Herculean self-sufficiency, but as *social* grit, built and sustained by others who invest in our wellbeing. Military cadres are social units that, at their best, do that job well. They need to support grit not only in war, but after war.

Seneca's Plea for Mercy

The core idea of having erred and finding a way forward as moral aspirant is nowhere more evident than in Seneca's essay *On Mercy*, addressed to Nero. In writing it, Seneca says he holds up "a mirror" for Nero to better see his ways. But the mirror is one Seneca holds up to himself, as well. As public spokesperson for the court, Seneca is also expressing the hope of the greater public that the tyrant will somehow show restraint, especially in the wake of having just murdered his half-brother Britannicus (at age 14) to thwart his claim to the throne. The essay's shadow twin is Seneca's play

[22] Lucius Annaeus Seneca, *Seneca: Six Tragedies*, Emily Wilson (ed.) (Oxford: Oxford University Press, 2010), "Hercules Furens", line 1265.

[23] Ibid., line 1275.

the *Trojan Women*. In the essay, we see the promise of mercy. In the play, we see the wasteland of a postwar world bereft of it.

Mercy, in the essay, is cast as the humane virtue in a world of human frailty. It is not pardon, "the remission of a deserved punishment," but "leniency in exacting punishment." It is "stopping short of what might have been deservedly imposed."[24] It is a staying of the hand, restraint and calming of vengeful anger, an acknowledgment that most accusers are not themselves free of all blame: "We have all sinned – some in serious, some in trivial things; some from deliberate intention, some by chance impulse, or because we were led away by the wickedness of others; some of us have not stood strongly enough by good resolutions."[25] Even if we become perfectly virtuous, we would have arrived there by some moral erring along the way. There is no sinless path in moral progress.

Seneca argues that mercy makes good on the gentler side of Stoicism, lost on critics who see only Stoicism's sturdy austerity. The Stoic moral tutor, like the good farmer (or vintner, as Seneca was in his country estates) is ready to adjust soil in need of nutrients, prop up a tree growing crooked, or prune the branches of another so those dwarfed by its shadow are now open to the light.[26] Moral tutelage, by analogy, is a matter of tender cultivation and willingness to show lenience, even if, by a stricter reading of rules and norms, harsh punishment might be rightfully imposed.

Such is Andromache's plea to Ulysses in the concluding scenes of Seneca's *Trojan Women*. The Greeks, despite their victory, find themselves at the end of this war, as at the beginning, stuck without the right winds to set sail. And following a familiar script, Calchas, the Greek priest, recommends Hector and Andromache's young baby son, Astyanax, be sacrificed and, too, that Polyxena, the young daughter of Priam and Hecuba, be slaughtered as a war bride on Achilles's tomb by his proxy, his son Pyrrhus. The children must bear the crimes of their ancestors. The ghost of Achilles kills his young bride and a baby boy is thwarted from becoming a warrior who can reignite another cycle of the Trojan War.

The future Trojan warrior boy must face his fate. But his mother, Andromache, is in a mortal battle with Ulysses to protect her innocent child. She has hidden him in her husband Hector's tomb, a sacred monument safe from enemy destruction. She begs Ulysses for mercy, for kindness, for herself as a hostage of war and as a mother whose child is her

24 Seneca, *On Mercy*, J. W. Basore (ed.), (Cambridge, MA: Harvard University Press, 1985), vol. I., 2.6.3–7.1.

25 Ibid., 1.7.

26 Ibid., 2.4.4; 2.7.4–5.

only comfort. The boy is no threat, she pleads, too young and without any power or backing to rearm a city. The boy may be a royal, but he is as good as an enslaved child now: Just put a yoke on his "royal neck."[27]

To kill him is a crime of war, protests Andromache, and the atrocity will be pinned not on the gods but on you, Ulysses. But a Greek warrior, set on vengeance, cannot stay the impulse. Anger, as Seneca so often teaches, can't be stopped, once set in motion: "I wish I could be merciful. I cannot," answers Ulysses.[28] The transmission of war, across generations, will go on, in violation of war's permissions. Once the appetite for warrior anger is whet, it knows no bounds. Innocent children, killed by mistake or vengeance, are war's pawns. Ulysses, ever the wily warrior and cunning strategist, can't find a strategy for showing mercy once the warrior impulse grabs hold.

What we next see is the stunning consequence of warrior rage: a little boy, Astyanyx, forced to step off the steep embankment that was once the site of his grandfather Priam's watch tower. The boy's body shatters with the impact of the plunge. His corpse is mangled, his skull cracks open, brains spurt out, a little boy pulverized as if by a high-impact bomb. Pleas of leniency, entreaties to restrain a victor's revenge, reminders that these children are victims, not contributors to war, the impotence of a ghost warrior groom, all reminders that though the aggression of the war is over, none of this stays the hand of ruthless rage. The afterwar can rage ruthlessly, however innocent its victims.

This is a strange play, we might think, for a Stoic moralist of calm. Or maybe not, for it's a cautionary tale about excessive punishment and the difficulty of staying the impulse of raging revenge in war. But it is also about leniency in the face of overzealous punishment, whether directed at an external enemy or the enemy within. We hope Ulysses might hear the plea for mercy, for the sake of a mother and a child. But also, for the sake of himself and his troops. For maybe he will come to feel guilt, and his soldiers, too, for what they have done to an innocent. Maybe the guilt will rack them for years to come and be the next feared Trojan War, but this one an inner war, that is fought over and over and over.

Mercy is, of course, far more elusive than anger. It requires discipline: first you have to vanquish anger, so that there is space to heal. Guilt is self-anger. Self-mercy, or perhaps, better, self-empathy, may be its therapy. But it needs others to be nurtured and sustained. That is Seneca's teaching.

[27] Seneca, *Seneca: Six Tragedies*, "Trojan Women", line 748.
[28] Ibid., line 764.

A Plea for Self-Empathy

Let's return a final time to the naval aviator Layne McDowell. There were no formal investigations of the incident over Kosovo. It is not clear if, by following fuller procedures, McDowell could have averted the accident. What we know is that McDowell, above all else, is his own judge and revisits the scene in flashbacks. He probably also revisits the scenes in his mind as he checks for accuracy in Chivers's typed-up account of his quotes and then later, his story, as Chivers narrates it, when he turns to reading the book.

What McDowell pictures in his flashbacks is strikingly like what Seneca pictures in Astyanax's demise – a young boy's body shattered, the back of his head missing, an innocent made all too vulnerable in war. What we hope for on behalf of this navy pilot is some mitigation of the self-punishment, some leniency and self-empathy that allows for a way to move beyond the rage of distress without losing the moral meaning that comes from the anguish. What we hope for is self-compassion. We want him to be able to loosen the rage of his felt guilt by imagining if he would blame others as harshly as he blames himself. His feelings may be apt, but relentlessly harsh. A kind of therapeutic temperance (in the name of psychic health) is required as a side constraint.

There is often asymmetry in how we hold self and others to account, especially, in the case of military moral injury: self-blame may be far harsher than blame for those who cover your back. But that opens a path for healing: We need to show ourselves the compassion we would show others in similar circumstances. Or show ourselves what we imagine that they show us. The benevolence of the benevolent spectator, at times, needs to become part of the moral self.[29]

There is a political lesson here, too. We who don't go to war need to start taking greater responsibility for the wars to which we send others to fight on our behalf. We need to carry the moral burden by doing a far

[29] This is not lost on clinicians working on therapeutic moral repair, especially in a technique called "adaptive disclosure" which focuses on a benevolent other from whom one can model self-benevolence and self-forgiveness. See Litz et al., *Adaptive Disclosure*; Brandon Griffin et al, "Development of the Self-Forgiveness Dual-Process Scale," *Journal of Counseling Psychology*, 65, no. 6 (November 2018), 715–26; Brandon Griffin et al., "Forgiveness as a Mechanism of Repair following Military-Related Moral Injury," in J. Currier, J. Nieuwsma, and K. Drescher (eds.), *Addressing Moral Injury in Clinical Practice* (Washington, DC: APA Publishing, 2019), 71–86; and Maguen et al., "Killing and Latent Classes." I am grateful to Brett Litz, Bill Nash, Shira Maguen, Brandon Griffin, and Natalie Purcell for conversation on moral injury and self-forgiveness and self-compassion over the years. I am especially grateful to Shira Maguen at the San Francisco VA for ongoing conversation about moral injury and repair.

better job than we have in the past two decades of knowing which causes of war are just and worth our nation's most precious resources.

Seneca is a complex spokesperson for moral healing and equanimity. What sometimes parades as calm conscience is the rumbling of the unconscious and its conflicts. He yearns for simplicity and tranquility, at the same time he is attracted to the messy world of high-stakes power and hierarchy. Modern-day warriors also live in complicated moral worlds – committed to excellence but working in institutions that vastly limit their individual control, exposed to situations that constantly test their best judgment and capacity for steady restraint. Exposure to moral injury is no surprise in these environments. But the Senecan lesson I have been urging is that moral injury may open the way for moral aspiration and repair. To read Stoicism as foreswearing the possibility of "good" moral distress is to miss Seneca's more profound lessons for moral resilience.[30]

[30] I am deeply grateful to Graham Parsons for all his work in organizing the West Point conference in October 2019 that brought us together and led to this book. It was an occasion to truly learn from others and share across disciplines. And I am very grateful to Mark Wilson for his keen editorial comments on the chapter in preparation for publication. I went on to develop some of the themes of the chapter in *Stoic Wisdom*.

CHAPTER 4

Political Humiliation and the Sense of Replacement

*Nir Eisikovits**

A great deal of international and domestic conflict is driven by feelings of humiliation. But what does political humiliation consist of? In this chapter, I argue that a key part of political humiliation involves the sense of being replaced. After looking at several case studies, including the rhetoric used by ISIS recruiters, the rise of revanchist Russian and Chinese foreign policies, and the language deployed by white supremacists to frame and justify their grievances, I point to some features of the sense of replacement. These include the loss of status and perceived break of a promise or denial of an entitlement – both coupled with a reactionary brand of nostalgia that aims to return the world to how things were before these losses occurred. The first section introduces the idea of humiliation in international and domestic conflict. The second uses four case studies to suggest that replacement is key in making sense of political humiliation. The third section offers several reasons why it is particularly important to understand the dynamics of replacement.

Humiliation in Politics

Recruitment materials for the Islamic State (ISIS) are rife with references to humiliations Sunnis have suffered at the hands of Shiites and Western powers. Video footage taken by a Vice News team that was allowed to accompany ISIS militants in 2014 depicts them destroying the Sykes-Picot era border between Syria and Iraq – long a symbol of humiliation of Arabs by colonial powers.[1] The lucky few who join ISIS, so its leaders promise,

* I'm grateful to Jake Burley, Nicholas Roberts, and Alan Zebek for help with research on white supremacy and "Great Replacement" theorizing, recent Chinese and Russian foreign policy, and the sense of replacement in interpersonal settings. I am indebted to Jeremy Wanderer for a series of illuminating discussions and critiques of the main argument of this chapter.

[1] *Vice News*, "Bulldozing the Border Between Iraq and Syria: The Islamic State," online video clip (August 14, 2014), www.youtube.com/watch?v=TxX_THjtXOw.

will help reverse that trend, as the group revives the historical Caliphate and returns a sense of virility and significance to the Muslim nation or *Umma*.[2]

Years earlier, reflecting on the surprise 1973 Egyptian-Syrian attack on Israel, Egypt's President Anwar Sadat explained that the motivation for the Yom Kippur War (or, as the Egyptians call it, the October War) was to reverse the humiliation Israel inflicted on Arab armies in 1967.[3]

Sweeping northeast over the map to Russia, much of that country's aggressive foreign policy in recent years – from the annexation of Crimea to President Vladimir Putin's insistence on playing a key role in Syria – can be traced to Russians' perception that their country was relegated to second-rate power status after the collapse of the Soviet Union and that the Americans and Europeans no longer took it seriously after 1990.[4]

China's energetic rise in international relations and its ambitious recent economic and infrastructure projects, including the launch of the Asian Infrastructure Investment Bank, the One Belt, One Road initiative, and the push to provide the world's 5G infrastructure, are best understood through (and have been justified by) a desire to emerge from the "century of humiliation" – the long period during which China's interests and its honor were compromised by foreigners. The Chinese Communist Party (CCP) explains and "sells" these massive projects to its own people as part of a move from political humiliation back to the days of the "middle kingdom," when China could claim to be at the heart of world affairs.[5]

Finally, a sense of humiliation has been part of the rhetoric that justifies white nationalism in the United States and Europe. In both regions, white nationalist's political messaging speaks of a once-ascendant, proud, nationally and culturally cohesive white culture that is being undone by

[2] See Gerges A. Fawaz, *ISIS: A History*, revised edn. (Princeton, NJ: Princeton University Press, 2017). For a focus on the role of humiliation in ISIS propaganda, see Nir Eisikovits, "ISIS, Humiliation and Political Philosophy," *Journal of Military Ethics*, forthcoming.

[3] Anwar Sadat, *In Search of Identity: An Autobiography* (New York: Harper Collins, 1978). Sadat writes on page 215: "the basic task was to wipe out the disgrace and humiliation that followed from the 1967 defeat. I reckoned it would be 1000 times more honorable for us…to be buried crossing the [Suez] Canal than to accept disgrace and humiliation."

[4] See, generally: Bobo Lo, *Russia and the New World Disorder* (Washington, DC: Brookings Institution Press, 2015); and Shaun Walker, *The Long Hangover: Putin's New Russia and the Ghosts of the Past* (New York: Oxford University Press, 2019).

[5] See, generally: Eric Anderson, *China Restored: The Middle Kingdom Looks to 2020 and Beyond* (Santa Barbara: Praeger, 2010); Ye Zicheng, *Inside China's Grand Strategy* (Lexington, KY: The University Press of Kentucky, 2011); Zheng Wang, *Never Forget National Humiliation: Historical Memory in Chinese Politics and Foreign Relations* (New York: Columbia University Press, 2014).

nonwhite, non-Christian foreigners, "invaders" who are depriving the "indigenous" whites of income, work, status, and ultimately their very culture.[6]

As these examples suggest, feelings of humiliation fuel a good deal of international and domestic conflict (let us put aside, for the moment, the important question of how to figure out when the feeling of humiliation is justified and when it is manufactured). And yet, political humiliation has received little philosophical scrutiny.[7] What does it mean to be politically humiliated? What are some of the special features of humiliation in international relations? In recent work[8] I have argued, contrary to scholars like Axel Honneth and Avishai Margalit, that there is more to political humiliation than misrecognition or the denial of one's human dignity. Humiliation and recognition are not different sides of the same coin; to be humiliated is not simply to be disrespected in the Kantian sense of that term. After all, we can feel humiliated by those whose recognition we don't seek or care about and we can feel severely humiliated even when our status as rational members of the human community is fully recognized or when others have no intention of questioning it. I further offered an account of humiliation in foreign policy and suggested that a humiliating foreign policy typically involves one or more of the following tendencies. First is the tendency of acting toward others while assuming their inability to respond, imposing one's will, knowing that one cannot be counteracted. Thucydides' Melian Dialogue provides the paradigm case: The Athenians decided to overrun the small island of Melos because they thought it was in their interest to do so and because they knew the Melians' objections were immaterial and impotent. A second feature typical of a humiliating foreign policy is to disregard the national heritage or history of others. Such a disposition was on display when the Americans allowed the Iraqi National Museum and

[6] See, generally, Francis Fukuyama, *Identity: The Demand for Dignity and the Politics of Resentment* (New York: Farrar, Straus and Giroux, 2018).

[7] Existing works are primarily in the fields of history, social psychology, and political science. For an outstanding recent historical survey of the role of humiliation in international affairs, see Ute Frevert, *The Politics of Humiliation* (New York: Oxford University Press, 2020). Evelin Lindner's important works in the psychology of humiliation are also worth noting: *Making Enemies: Humiliation and International Conflict* (Santa Barbara, CA: Praeger, 2006) and *Honor, Humiliation, and Terror: An Explosive Mix and How to Diffuse it with Dignity* (New York: Dignity Press, 2017). Finally, see Deepak Tripathi, *Imperial Designs: War, Humiliation & the Making of History* (Washington, DC: Potomac Books, 2013); Bertrand Badie, *Humiliation in International Relations: A Pathology of Contemporary International Systems* (Oxford: Hart Publishing, 2017); and Joslyn Barnhart, *The Consequences of Humiliation: Anger and Status in World Politics* (New York: Cornell University Press, 2020) for further recent treatments.

[8] Eisikovits, "ISIS, Humiliation, and Political Philosophy."

National Library to be looted while making sure to secure the country's Oil Ministry during the early days of the 2003 occupation. Third and finally, political humiliation often manifests as the tendency to dismiss and ignore the complex social and cultural realities making up the lives of those we come into contact with. The powers that divided the "Levant" after World War I were ignorant of the intricacies of ethnic and tribal tensions in the area, and sliced it up on their maps as if these realities did not exist.

But the parsing of humiliation from failures of recognition and even the characterization of what a humiliating foreign policy entails do not tell us what it feels like to be humiliated. In the rest of this chapter, I will focus on the contents of that feeling and, in particular, on the role the sense of replacement plays in political humiliation.

The Sense of Replacement

A key feature of the phenomenology of humiliation – in international relations, domestic strife, and even intimate relationships – has to do with the sense of being replaced, with anxieties and frustrations about replacement. To make the case for this, let's take a closer look at some of the examples discussed earlier.[9]

ISIS propaganda is replete with pronouncements about the lost glory of the *Umma* or Muslim nation – about honor lost to crusaders and other "infidels." Muslims, so the story goes, used to be warlike and widely revered, but have surrendered their ferocity and pride to other powers. Muslims used to be ascendant in war and culture but now they have become subservient to the west. ISIS promises to change these trends. *The Management of Savagery*, an Al Qaeda handbook widely distributed among ISIS recruits, describes the transition of Muslims from "domesticated" passivity to newfound agency:

> Those who have not boldly entered wars during their lifetimes do not understand the role of violence and coarseness against the infidels in combat and media battles. The stage of domesticating the Muslims which they have already passed through has had an effect on them. The reality of this

[9] I'm not claiming that the sense of being replaced is a sufficient or even necessary condition for the feeling of humiliation to set in. It is certainly possible to be politically humiliated without feeling personally or collectively displaced. The Germans felt humiliated at the 1919 Paris Conference not because the Americans, British, and French somehow "replaced" them on the world stage, but because their interests were not taken seriously and a disproportionate degree of responsibility for the war's damage was imposed on them. I am, however, arguing that as a matter of political reality, in political practice and rhetoric, the feeling of being replaced is very often interchangeable with the feeling of being humiliated.

> role must be understood by explaining it to the youth who want to fight. They are different from the Arabs at the beginning of the Prophet's mission. The Arabs used to fight and know the nature of wars.[10]

Writing in the first issue of *Dabiq*, the group's promotional magazine, ISIS spokesperson Al Adnani proclaims:

> The time has come for those generations that were drowning in oceans of disgrace, being nursed on the milk of humiliation, and being ruled by the vilest of all people, after their long slumber in the darkness of neglect—the time has come for them to rise. The time has come for the ummah of Muhammad…to wake up from its sleep, remove the garments of dishonor, and shake off the dust of humiliation and disgrace, for the era of lamenting and moaning has gone, and the dawn of honor has emerged anew.

Examples from ISIS propaganda can be multiplied but the theme is consistent: the organization is there to return Muslims to a position of power in world affairs – a position from which they were displaced in the last few centuries. The fundamentalist tendencies of ISIS – the literal interpretation of religious texts, the insistence on the validity of seventh-century practices (sexual slavery, decapitating or burning prisoners alive, a dress code that emulates the way people wore their clothes during the time of the prophet), the desire to reinstate the pan-Islamic Caliphate – are not animated only by religious reasons. They are also expressive of a deep nostalgia and are part of an attempt to recreate a time when Sunnis were revered. A large part of ISIS's justifying narrative, then, is that Sunnis have been replaced as key holders of power and prestige by infidels – Shiites, Crusaders, Jews, and so on. The organization's goal is, quite literally, to turn back the clock on that replacement.

The sense of replacement is also key to understanding Russia's recent foreign policy. From the annexation of Crimea to its involvement in Syria, many of the country's actions on the international scene are best understood as responses to being marginalized and dismissed after the collapse of the Soviet Union. Indeed, Putin and other Kremlin officials consistently claim that Russia was relegated to second-rate power status after 1990, that it has been disrespected, discarded, and replaced on the world stage, and that they (and only they) can halt this decline. Some of the primary complaints of the Russians include the NATO intervention in Bosnia in 1995 against their wishes, the quick dismissal of Russia's offer to help prosecute

[10] For an excellent discussion of the centrality of this text for ISIS see chapter 1 of Fawaz, *ISIS*.

the "War on Terror" after the attacks of September 11 (specifically the offer to help with intelligence in Afghanistan), the beginning of the 2003 war in Iraq (again against Russia's wishes), and the second round of NATO expansion in 2004 that removed former Soviet Republics Estonia, Latvia, and Lithuania from Russia's traditional sphere of influence.

In a recent book, journalist Shaun Walker relays a phone conversation between U.S. President Bill Clinton and Russia's Boris Yeltsin in which the former ignored the latter's objections to a planned bombing campaign in Serbia. Putin, who learned of the exchange, was shocked to discover how easy it had become to dismiss Russia's interests. He vowed to change that.[11] Right before becoming president, Putin wrote the following in the newspaper *Nezavisimaya Gazeta*:

> for the first time in the past 200–300 years, Russia faces the real danger of being demoted to the second, or even the third tier of global powers. To prevent this from happening, we as a nation are to pull all our intellectual, physical, and normative resources, thus ensuring that ours remains a first-tier nation. Our efforts must be united, coherent and constructive, for no one else would do this work for us.[12]

Shortly after that, in an "Open Letter to Voters" published in *Izvestia*, *Kommersant*, and *Komsomolskaia Pravda* newspapers on February 25, 2000, Putin added:

> Our priority is to restore the personal dignity of the people in the name of the dignity of the nation…Russia has long ceased to be just a reduced map of the Soviet Union; it is a confident power with a great future and a great people…The past decade has brought dramatic changes to the consciousness of the people. Our citizens are not yet rich, but they are independent and self-confident…Our army is emerging from a prolonged crisis with honor and is becoming ever more efficient and professional….True, Russia has ceased to be an empire, but it has not wasted its potential as a great power…It is unreasonable to be afraid of a strong Russia, but it should be reckoned with. Insulting us is counterproductive.[13]

The anger about the diminution of Russia on the international scene, the resentment about its displacement as a great power, and the desire to see it reinstated comes across in interviews with leading Russian intellectuals and

[11] Walker, *The Long Hangover*. See also Shaun Walker, "The Humiliation that Pushed Putin to Try and Recapture Russian Glory," History.com, February 23, 2018. www.history.com/news/vladimir-putin-russia-power.

[12] Vladimir Putin, "Russia at the Turn of the Millennium," *Nezavisimaia Gazeta*, trans. Alan Zebek (December 30, 1999).

[13] Vladimir Putin, "Open Letter to Voters," Kremlin.ru (February 25, 2000) http://en.kremlin.ru/events/president/transcripts/24144.

journalists as well. Here is well-known director and screenwriter Andrei Konchalovsky in an interview with *Vesti Nedeli* on January 24, 2009:

> Clinton was good to Russia, because Russia did not exist on the map. Clinton was good to Russia because Russia had kneeled before America for it was living on American money for a decade. Today's Russia is no longer kneeling, and has the means to defend its interests on the global stage. No one likes it, but they have to deal with it. I do not think that Russia and the US can be friends. Nevertheless, their relationship can be mutually respectful. Russia's interests must be acknowledged and accepted as given, as the status quo.[14]

And, more to the point, the prominent journalist Vladimir Pozner had the following to say in an interview with the newspaper *Argumenty i Fakty* on April 02, 2015:

> [Imperial ambitions] are like a phantom-limb pain. When Western states tell us, "You are not great...You are a second- or even fifth-tier nation [and] your time has passed..." it only adds to those conscious and subconscious grievances over the lost empire. Yet, people still deem themselves as part of a great power. Telling them that it is no longer the case, only exacerbates their sense of loss...[15]

Chinese foreign policy is similarly responsive to a long history of perceived humiliations. China's so-called "Century of Humiliation" refers to a period from the mid-nineteenth to the mid-twentieth century, during which the country suffered a series of defeats in military engagements and was subjected to numerous disadvantageous treaties at the hands of Britain, France, Russia, and Japan. The Chinese Communist Party (CCP) has used "the Century of Humiliation" to offer a narrative of Chinese decline – from the time of the "Middle Kingdom" (a name coined to denote China's pivotal and superior role in international affairs under the Qing Dynasty) to a marginal power dismissed by neighbors and Western powers alike. And, naturally, the role of the CCP – like that of ISIS and Putin in their own regional contexts – would be to restore China to its past glory.[16] In the aftermath of the deadly suppression of protests in Tiananmen Square,

[14] Andrei Konchalovsky, interview with *Vesti Nedeli* (January 24, 2009) www.vesti.ru/article/2162310. Translation courtesy of Alan Zebek.

[15] Vladimir Pozner, interview with *Argumenty i Fakty*, (February 4, 2015), https://aif.ru/politics/russia/1438503. Translation courtesy of Alan Zebek.

[16] See, Anderson, *China Restored*; Zicheng, *Inside China's Grand Strategy*; and Zheng, *Never Forget National Humiliation*. See also the Shivshankar Menon, "How China Bucked Western Expectations and What it Means for World Order," Brookings Order from Chaos blog (March 10, 2016), https://brook.gs/3vZc8al.

the CCP began the "Patriotic Education Campaign." The campaign aimed to educate China's youth about the country's humiliating experience at the hands of Western and Japanese aggressors and domestic "traitors," and to show them how "the CCP-led revolution had changed China's fate and won national independence."[17]

According to the most updated version of this narrative, China is now on the verge of reclaiming its rightful position in the world. Owing to Xi Jinping's emphasis on restoring national greatness, party leaders are portrayed as heirs to the dynasties of the eighteenth century, when Qing emperors demanded and received respect and even supplication from Western envoys.[18] This push makes sense of the formation of the Asian Infrastructure Investment Bank which allows China to take its fiscal future (and that of some of its neighbors) into its hands rather than depend on the Asian Development Bank, the World Bank, or other monetary institutions that are shaped by the United States. It similarly underpins the Belt and Road initiative, also known as "One Belt, One Road" (OBOR), an ambitious infrastructure project covering much of the region and supporting about a third of the world's GDP.[19]

Finally, White supremacists in the United States as well as their nativist, anti-migration counterparts in Europe speak incessantly of the fear and humiliation of being displaced and replaced. "Jews will not Replace Us," the neo-Nazis famously chanted in Charlottesville in August of 2017. That chant finds its inspiration in a book by far-right French writer, Renaud Camus. The influence of his *Great Replacement* has been wide-ranging.[20] As James McAuley recently wrote in *The Nation*:

> the term "great replacement" has become a rallying cry of white supremacists around the world – the demonstrators who stormed through Charlottesville, Virginia, in August 2017; the man who killed 11 worshippers at the

[17] Zheng, *Never Forget National Humiliation*, p. 116.

[18] See Frevert, *The Politics of Humiliation*, chapter 3.

[19] Christina Müller-Markus, "One Belt, One Road: The Chinese Dream and Its Impact on Europe," *Notes Internacionals CIDOB*, 148 (2016), 1–6; Hong Yu, "Motivation behind China's 'One Belt, One Road' Initiatives and Establishment of the Asian Infrastructure Investment Bank," *Journal of Contemporary China*, 26, no. 105 (2017), 353–68.

[20] Renaud Camus, *Le Grand Remplacement* (Plieux : Chez l'auteur, 2019). The English version is titled *You Will Not Replace Us!* and was published in 2018 by the same publisher. For background on Camus see, Simon Murdoch and Joe Mulhall, *From Banners to Bullets: The International Identitarian Movement* (London: HOPE not hate Charitable Trust, 2019), https://bit.ly/3QgXK5t. See also Cécile Leconte, "The Socio-Political Career of the Expression 'the Great Replacement' Among Right-Wing Party Networks in Germany: The Case of the Alternative for Germany (AfD) Party," *Politix*, 126, no. 2 (2019), 111–34.

> Tree of Life synagogue in Pittsburgh in October 2018; and especially Brenton Tarrant, who perpetrated the New Zealand mosque attacks in March of 2019. Tarrant posted his own "The Great Replacement" – a 74-page online manifesto – before murdering 51 people.[21]

Camus claims that a country's national identity is a function of the shared cultural norms and heritage embraced by its population. Based on the (unsupported) assumption that such coherence and uniformity really exist or ever existed, Camus postulates that if another population is introduced into such a coherent context under circumstances that overwhelm the native population's growth rate, the identity or essence of the existing civilization will be replaced by that of the incoming one. The argument's second premise (in addition to the existence of "cultural essences") is that people – in their individual or collective capacities – resent, indeed hate, feeling replaced. As Camus put it in a 2017 interview with *Vox*'s Sarah Wildman, "the refusal to be replaced is a very strong feeling in man. It doesn't really need to be put into hearts and into minds."[22]

The sense of replacement has elicited wildly different responses from the actors we have surveyed. ISIS responded with murderous brutality against anyone who does not meet its standard of Sunni purity. White supremacists use it to justify their xenophobia and racism and to throw their support behind nativist politicians. China and Russia have, in different ways, used the sense of replacement to justify and sell their populations on more aggressive foreign policies.

But though responses to the sense of replacement vary, there are common themes in how it operates. In all cases, there is a feeling of lost status, a perception of lost ascendancy, of being removed from importance and consequence – usually by someone (or in favor of someone) not viewed as a worthy competitor. China was once the "middle kingdom" in the heart of world affairs, but the Century of Humiliation pushed it aside. Russia was once a superpower, first in the race to space, a formidable competitor with the United States for nuclear supremacy, and yet, since 1990, it has been relegated to the minor leagues of international relations. White men in the United States and Europe once enjoyed unquestioned ascendency, but now they see people of different races, ethnicities, and genders gain influence and prominence. The Islamic Caliphate used to reign supreme

[21] James McAuley, "How Gay Icon Renaud Camus Became the Ideologue of White Supremacy," *The Nation* (June 17, 2019), https://bit.ly/3zQOHB3.

[22] Sarah Wildman, "'You will not replace us': a French philosopher explains the Charlottesville rant," *Vox*, August 15, 2017. https://bit.ly/3QDaDq6.

across much of the world and Muslims used to lead in both military and cultural pursuits while Europe was a backwater. Now the power relation has been reversed. Relatedly, in each case there is a sense of entitlement denied, of a promise – explicit or implicit in political culture – broken, an identity-defining, long counted-on birthright withdrawn. For the Russians and Chinese, it was the promise that they would always be citizens of great nations, a greatness that would compensate for and make sense of the sacrifices they were often called to make for the state. For white supremacists both in the U.S. and elsewhere, it was that white men would always hold sway, that they would have the most influence in how their countries would be ruled.

The sense of replacement, then, is about the diminution of status, the loss of an entitlement to someone perceived as unworthy of it, and, importantly, it is also about the desire to take these back. Surrendering to the sense of replacement is not simply sinking into a politics of nostalgia. It is, rather, taking on an *actionable* nostalgia involving the desire to recreate the specialness, to bring back the ascendency, to reclaim the stolen birthright. In other words, the story of political humiliation, in so far as it has to do with replacement, is often a story of being removed from a pedestal, losing the assumed, natural, taken-for-granted preeminence that was constitutive of one's very political identity. It is the story of watching someone else instilled in your place – someone who looks and speaks differently, someone not from here, someone who you perhaps were accustomed to historically look down upon (as the Russians dismissed the "materialist" Americans, and as the Chinese dismissed the "barbarian" Japanese and westerners), and the angry determination to win it back and return things to their "natural" order.

Finally, the fact of being replaced – on the world stage or as part of the elite in one's country – gives one a sense that one is replaceable, disposable, worthless; it gives the lie to the rhetoric of political specialness, manifest destiny, chosen-ness, and exceptionalism that so many of the political players we have discussed like to instill. And it's the recognition of being replaceable or interchangeable that is humiliating – perhaps even more than the actual loss of entitlements and exclusive access to social and material goods.

Why We Need to Understand the Sense of Replacement

There is, of course, much more to political humiliation than the sense of being replaced. As mentioned above, among other features, political humiliation consists of the feeling of being powerless against the will of others, of not being a worthy subject of promises, of mattering for purely instrumental reasons. Why, then, focus on the sense of replacement? Why is the feeling of replacement particularly interesting philosophically? Why it is important, politically, to understand it better? In what follows I sketch four answers.

First, the sense of replacement crosses the divide between the personal and political. The self-same feeling of replacement that drives interpersonal humiliation is present in politics. The betrayed spouse jettisoned for another lover feels (among other things) that they have been replaced, discarded, that what they thought was a unique status in the regard of their partner was not unique after all. They are disposable rather than "the one." And that sense of replacement is the source of a great deal of anger. The same dynamics apply with varying degrees of intensity in nonromantic friendships, when, for example, an old friend has been discarded or neglected in favor of a new one.[23]

The similarity to the sense of replacement reported in the political settings we discussed is striking. What are we to make of this similarity? What does it mean that the phenomenology of replacement is so similar across interpersonal and political contexts? For one thing, it may suggest something about the intensity and primacy of the sense of replacement: that this feeling travels unchanged between intimate and political settings might attest to its unique force. There are other psychological dispositions that are salient in both interpersonal and political settings. But in many of these cases, the transition from the personal to the political takes on a metaphorical aspect: individuals can heal from an injury, and we also speak of "national healing," but what, exactly, does it mean for a nation to heal from an episode of mass atrocity or historical injustice?

Similarly, individuals can forgive offenses perpetrated against them, or reconcile with those who hurt them and, of course, we often speak about political forgiveness and reconciliation (especially after Archbishop Tutu framed and justified the work of the South African Truth and Reconciliation Commission as an instrument to promote national forgiveness and reconciliation), but what does it mean for one group to forgive another? Reconcile with each other? Do groups really heal, forgive, and

[23] On replacement in interpersonal relationships, see John Tooby and Leda Cosmides, "Friendship and the Banker's Paradox: Other Pathways to the Evolution of Adaptations for Altruism," in W. G. Runciman, John Maynard Smith, and R. I. M. Dunbar (eds.), *Evolution of Social Behaviour Patterns in Primates and Man* (Oxford and New York: Oxford University Press, 1996), 119–43; and J. A. Krems et al., "Friendship Jealousy: One Tool for Maintaining Friendships in the Face of Third-Party Threats?" *Journal of Personality and Social Psychology*, 120, no. 4 (2021), 977–1012.

reconcile in the same way individuals do? I am not arguing that it is impossible for polities to heal, forgive, or reconcile, and these dispositions are, of course, the subject of a rich psychological and philosophical literature.[24] I am, rather, making the more modest claim that in the transition from the psychological to the political, these terms lose some of their clarity, that their application becomes fuzzier in a way that does not hold when it comes to the sense of replacement.

The phenomenology of the personal and political versions of the sense of replacement and the resulting anger, disappointment, and resentment appear remarkably similar. In addition to pointing to the intensity of the sentiment itself, these similarities suggest that research on interpersonal feelings of replacement may be directly instructive for understanding the political manifestation of the feeling (and vice versa) – more so than in the case of other emotions. But there's an important caveat: though the feelings may be almost identical, the grounds for the feelings are not. While the sense of replacement and the resulting humiliation between intimate partners is often well founded, and while the scorned partner often has a moral claim on not being replaced (the explicit or implicit commitment to loyalty, for example, in a committed monogamous relationship), the sense of replacement in politics is often both empirically false and morally groundless. American white men, for example, are not really losing their ascendency – their sense of grievance is largely manufactured by political manipulation.[25] And even if they were losing their ascendency, in political society, unlike in a marriage, citizens are not and should not be promised a unique preferential status.

Second, and closely related to this last observation, a focus on the sense of replacement helps us distinguish normative and descriptive aspects of political humiliation. Groups certainly have a normative claim against being humiliated – for example when they are simply acted upon because they are powerless to resist, or when they are dismissed as subjects of binding promises, or when their cultural heritage is ignored. The Arab tribes

[24] See, for example, Charles Griswold, *Forgiveness: A Philosophical Exploration* (New York: Cambridge University Press, 2007); Colleen Murphy, *A Moral Theory of Political Reconciliation* (Cambridge: Cambridge University Press, 2010); Nir Eisikovits, *Sympathizing with the Enemy: Reconciliation, Transitional Justice, Negotiation* (Boston, MA: Brill, 2009); Nancy Sherman, *Afterwar: Healing the Moral Wounds of Our Soldiers* (New York: Oxford University Press, 2015).

[25] As of 2017, whites comprised 73 percent of the U.S. population, and as of 2019, Christians comprised 65 percent. See United States Census Bureau, "Population, Population Change, and Estimated Components of Population Change: April 1, 2010 to July 1, 2020," Census.gov, https://bit.ly/3QvoLyY.

who were cheated out of British and French promises for self-determination were correct to feel humiliated. The same is true of Palestinians who were informed by Golda Meir that "there [is] no such thing as Palestinians,"[26] or of Iraqis who realized that the American security forces were far more concerned about defending their Oil Ministry than national museum. But Russia does not have a moral right to keep its status as a great power, China does not have a moral right to remain the "middle kingdom" in perpetuity, and white men do not have a right to lasting social prestige and influence in the United States and Europe. The sense of replacement in all of these cases is probably psychologically real, the resulting humiliation is often deeply felt, and it is potentially explosive, but it is not morally justified – it does not arise from nor does it generate a legitimate moral claim.

Stated differently, there is often a difference between the moral status and the psychological intensity of the sense of replacement in politics. That sense does not necessarily generate a moral claim because, in politics, we do not normally have any political entitlements to remain in a position of ascendency or to be held in unique regard. And yet feelings of replacement, even when they are not justified, are real and can signal real trouble to come. In international politics as well as in domestic strife, the rise of a sense of replacement, regardless of its moral status, is predictive of instability and violence. Complaints about political replacement can function, then, as an early alert system, a seismograph, if you will, foretelling political turbulence.[27]

Third, understanding the sense of replacement can help us distinguish between different varieties of conservatism. In particular, it can help us to tell apart legitimate political conservatism from the fundamentalist, dangerous variety. In a recent study, Edmund Fawcett distinguishes between the "right" and the "hard right."[28] The former is a moderate conservatism skeptical of grand social schemes and more protective of lasting and meaningful features of the *status quo* than averse to change itself. The "hard right," on the other hand, is reactionary and nostalgic – gripped with a

[26] Quoted in Baruch Kimmerling and Joel Migdal, *The Palestinian People: A History* (Cambridge, MA: Harvard University Press, 2003), p. xxvi.

[27] It is interesting, but beyond our purview, to apply this predictive tool to another context all together – automation and the changing nature of work. According to many estimates, automation is going to replace and displace between a quarter and a third of the workforce in the next thirty years. In large chain stores, we are already seeing many employees working alongside the machines that will ultimately replace them. These trends are likely to bring about feelings of replacement on an unimaginable scale and, if the preceding analysis is applicable, a great deal of political unrest.

[28] Edmund Fawcett, *Conservatism: The Fight for a Tradition* (Princeton, NJ: Princeton University Press, 2020).

profound anxiety about the loss of status. The project of the hard right is not to counter-revolution with reform, not to reinterpret a tradition in a way that makes it cohere with current developments, but rather to resist all change that challenges the status of certain elites. Its aim is to retrench, and, if perceived loss of status has occurred, to return to a mythically pure and innocent time that preceded the loss. For Fawcett, the history of conservatism has as much to do with the tension between these two strands as it does with the struggle against liberalism and progressivism.

A focus on the sense of replacement is helpful in distinguishing these two brands of conservatism and indicating when we are dealing with the nostalgic, reactionary variety. As the case studies we have discussed suggest, the feeling of being replaced is usually accompanied by a strong desire or political ambition to turn back the clock, to return to Eden, to go back to how things were before standing was lost and disappointment set in. The presence of the sense of replacement is thus indicative of the reactionary conservative impulse – the virulent form of conservatism that seeks not simply to ease the impact of social changes and make sure key traditions survive it, but to subvert any kind of alteration, to frame all change as a zero-sum game which robs people of key features of their identity. White supremacists, Russian and Chinese nationalists, ISIS leaders – in their various ways and to various degrees – do not want to make room in the new world order for their views. They want to revert back to the old-world order in which their views and way of life were uncontested.

Fourth, and finally, understanding the sense of replacement gives us useful insight into the so-called Thucydides Trap, according to which the rise of a new power leads to clashes between the established and emergent forces.[29] By some estimates, up to 75 percent of such competitions result in armed conflict.[30] The examination of the sense of replacement suggests that the anxiety stirred for the established power is not so much about the rise of the new actor or the new way of life it portends as it is about the loss of status that comes with it – the sense of decline, replacement, and, ultimately, humiliation that go with no longer "counting." Russia was initially willing to live with the fast-rising American hegemony, as long as its

[29] As Thucydides puts it in the beginning of the *History*: "The real cause, however, [for the breakout of war] I consider to be the one which was formally most kept out of sight. The growth of the power of Athens, and the alarm which this inspired in Sparta, made war inevitable." Thucydides, *The Landmark Thucydides*, Robert Strassler (ed.), (New York: Touchstone, 1998), book I, section 23.

[30] For an interesting application of the Thucydides Trap to the Chinese-American context, see Graham Allison, *Destined for War: Can America and China Escape the Thucydides Trap?* (New York: Mariner Books, 2017).

status as a significant power was maintained – or at least a semblance of it. Putin's revanchist push began not so much due to the exponential rise of American power but because of the dismissal of Russian significance. This insight both reverses the emphasis of the Thucydides Trap – the decline of our vigor rather than the rise of the newly vigorous is what makes us jittery – and offers a way out of it. If the rising power is clever and generous enough to rise in a way that allows the declining actor to either maintain some of its status or at least save face, some of the risk embodied in Thucydides' insight may be averted.

Conclusion

The sense of replacement is a key part of the phenomenology of political humiliation. That phenomenology is remarkably similar across intimate and political settings, and that similarity points to the force of the sentiment and to the relevance of literature on interpersonal feelings of replacement for better understanding political manifestations of the feeling. A sense of humiliation that is based primarily on being replaced does not typically raise normative claims, and yet it is dangerous. The fact that the sense of replacement in politics is not morally justified may tempt us to dismiss it, but that would be a mistake. Justified or not, the sense of replacement is predictive of violence and, as a result, it is in the best interest of policymakers to watch for it and take precautions when they spot it. More empirical work is merited on the factors that make the sense of replacement particularly dangerous. It would also be helpful to better understand what, if anything, can be done to reverse or defang the sense of replacement. Are there any methods of reintegration that can specifically target the sense of replacement? Do existing methods of deradicalization take that sense and its toxic legacy into account? Should they? These questions point to the need for further normative and social-scientific investigations into the role of replacement in politics.

CHAPTER 5

Minimum Moral Thresholds at War's End

Colleen Murphy

Just war theory provides normative prescriptions for the onset, conduct, ending, and aftermath of war. Transitional justice provides normative prescriptions for dealing with widespread human rights violations characteristically committed during conflict and/or periods of repression. In both cases, the ambitious transformative aspirations contained in just war theory and theories of transitional justice fail to be realized in practice. At the time of this writing, the U.S. military has withdrawn from Afghanistan, following an almost twenty-year period of engagement. The Taliban, which in part sparked the United States' engagement, now governs Afghanistan. Almost thirty years ago, South Africa became the global face of transitional justice in its transition from apartheid to multiracial democracy. Ongoing violence following the arrest of former President Jacob Zuma on charges of corruption speaks to the limits of transformation there. While the immediate cause of the protests is the arrest, the underlying motivation is the enduring legacy of racialized apartheid in the form of economic and social exclusion for the majority Black South African population.

Just war theory and transitional justice both provide normative prescriptions for profoundly nonideal circumstances. Indeed, realists in both cases simply deny that morality applies to these situations. Moral theorists, however, believe in the possibility of distinguishing morally defensible from morally unjustified conduct, even in the exigencies of war and transition. Yet a yawning gap remains between the normative picture of just war and of transitional justice on the one hand, and the descriptive reality of contemporary conflicts and transitional justice practices on the other. To engage with the reality of each practice in a way that will prove actionable for its participants, new forms of normative guidance are needed. Or so I argue in this chapter.

The first part of this chapter provides an overview of the relevant normative aspirations of just war theory and transitional justice. The second section focuses on the critical turn in scholarship in both just war theory and transitional justice. I identify two analogous points of emphasis in that scholarship, points which underpin similar normative challenges and normative complaints that both fields face. My interest in this section is in explaining some of the reasons that achieving the normative goals of just war theory and transitional justice has proven so difficult in practice, and the challenges so intractable. The third and final section suggests that to engage more closely with warfare as it exists and transitional justice as it occurs, thresholds of minimal normative success must be identified.

Just War Theory and Transitional Justice: A Brief Overview

Just War Theory

Just war theory provides a set of moral criteria governing the initiation, conduct, winding down, and aftermath of warfare.[1] Such criteria inform and are reflected in international law. They allow us, for instance, to distinguish humanitarian intervention from aggression (a crime against humanity), and justified airstrikes from indiscriminate bombings (a war crime). The exigencies of warfare, from the perspective of just war theory and international law, do not render morality irrelevant. *Jus ad bellum* specifies the kinds of reasons for which warfare can be legitimately undertaken (e.g., self-defense, humanitarian intervention), by whom, with what probability of success, and the steps that must be taken prior to the onset of conflict so that war remains a last resort. In sum, the *jus ad bellum* stipulates what, morally speaking, constitutes a good war.

Jus in bello delimits the kind of conduct that is legitimate in warfare. Guidelines specify who and what count as legitimate targets, as well as what amount and kind of force are permissible. Traditional accounts, and international law, treat the first two categories of just war theory as independent. Combatants are morally equal regardless of the justice of the side for which they fight. Noncombatants are never to be intentionally targeted.[2]

[1] Seth Lazar, "War," in Edward N. Zalta (ed.), *The Stanford Encyclopedia of Philosophy* (Spring 2020 edition).

[2] Revisionary accounts challenge the legitimacy of this separation and the normative basis for the prohibition on intentionally targeting noncombatants. In such revisionist accounts, combatants fighting a just cause are not necessarily liable to be killed, while noncombatants from the community pursuing an unjust war may sometimes be legitimately targeted. See, for example, Jeff McMahan, *Killing in War* (Oxford: Oxford University Press, 2009); Victor Tadros, *To Do, To Die, To Reason Why: Individual Ethics in War* (Oxford: Oxford University Press, 2020).

Jus ex bello provides guidelines for ending war.[3] Such guidelines are formulated in response to the recognition that a war justly undertaken may, as the course of the war proceeds, become unjust given the costs or importance of continuing. The converse considerations apply.

Jus post bellum principles govern the aftermath of war. These principles outline, for example, the kinds of surrender that may be permissibly sought by the winning side,[4] and the permissible means for dealing with violations of *jus ad bellum* and *jus in bello*.[5] Guidelines as to whether punishment is required as a matter of justice for such violations and by whom such punishment may be permissibly meted out are articulated. *Jus post bellum* also articulates criteria for just postwar reconstruction,[6] providing guidance as to who should bear the responsibility for reconstruction (e.g., the international community or parties to the conflict), how that responsibility should be distributed, and what the goal for reconstruction should be (e.g., the *status quo ante* or a normative standard).[7]

[3] Darrel Moellendorf, "*Jus ex Bello*," *Journal of Political Philosophy*, 16 (2008), 123–36.

[4] See Brian Orend, "Jus Post Bellum," *Journal of Social Philosophy*, 31, no. 1 (2000), 117–37; Brian Orend, *War and International Justice: A Kantian Perspective* (Waterloo, Ontario: Wilfrid Laurier University Press, 2000); Brian Orend, "Justice after War," *Ethics & International Affairs*, 16, no. 1 (2002), 43–56.

[5] Orend, "Justice after War"; Davida E. Kellogg, "Jus Post Bellum: The Importance of War Crimes Trials," *Parameters*, 32, no. 3 (2002), 87–99.

[6] Gary J. Bass, "Jus Post Bellum," *Philosophy and Public Affairs*, 32, no. 4 (2004), 384–412; Alex J. Bellamy, "The Responsibilities of Victory: *Jus Post Bellum* and the Just War," *Review of International Studies*, 34 (2008), 601–25; Mark Evans, "Moral Responsibilities and the Conflicting Demands of *Jus Post Bellum*," *Ethics & International Affairs*, 23, no. 2 (2009), 147–64; and Alexandra Gheciu and Jennifer Welsh, "The Imperative to Rebuild: Assessing the Normative Case for Postconflict Reconstruction," *Ethics & International Affairs*, 23, no. 2 (2009), 121–46.

[7] As with the relationship between *jus ad bellum* and *jus in bello*, the question of how to understand the relationship between, *jus ex bello*, *jus post bellum*, and the other parts of just war theory is a subject of ongoing debate. Some see the criteria of *jus post bellum* as continuous with other doctrines, where the aim of reconstruction is set by the reasons for which war was undertaken. If defense of sovereign territory is the reason for war, victors are limited by that aim in what they can do after war; a postwar land grab outside the scope of territory defended would be impermissible. However, other theorists expand the scope of action postwar beyond the reasons for which war was undertaken. On this point, see Bellamy, "The Responsibilities of Victory." The victor's role in the carnage of warfare can ground a greater set of postwar obligations, such as ensuring a modicum of stability in an enemy state through temporary occupation. See, for example, Evans, "Moral Responsibilities." The claim that the principles are interdependent is also challenged; this is necessary for *jus post bellum* to provide normative guidance for how actors involved in unjust wars (from the perspective of *jus ad bellum* for example) should act. See Bellamy, "The Responsibilities of Victory," 610. Michael Walzer, for example, appeals to democratic principles, specifically the goal of aiding defeated peoples to achieve self-determination. See Michael Walzer, "Just and Unjust Occupations," *Dissent*, 51, no. 1 (2004), 61–63.

Transitional Justice

Transitional justice is the process of dealing with widespread wrongdoing.[8] More than forty countries have explicitly pursued transitional justice in recent decades, characteristically as part of a process of transitioning away from extended conflict and/or repression to more peaceful and less oppressive forms of governance.[9] Prominent examples where transitional justice was pursued include the aftermath of the Balkan Wars following the break-up of Yugoslavia in the 1990s, the transition from apartheid to multiracial democracy in South Africa in the 1990s, and the ongoing process of implementing the 2016 Final Agreement to end more than fifty years of conflict between the Colombian government and the Revolutionary Armed Forces of Colombia (FARC).

To understand transitional justice, it is necessary to spell out the forms and normative goals of the process used as well as the feature of widespread wrongdoing that make a response normatively urgent. Transitional justice processes can take many different forms. Domestic, ad hoc, permanent international, and hybrid *criminal trials* hold perpetrators to account for wrongdoing. Hybrid criminal trials are half-international, half-national in origins, subject-matter jurisdiction, and/or personnel. *Truth commissions* are official, ad hoc bodies established with a mandate of documenting the causes, extent, and consequences of a delimited set of human rights violations.[10] Such documentation typically entails naming perpetrators responsible for and victims of such violations. *Amnesty* provisions remove the prospect of criminal or civil liability for specified perpetrators.[11] Eligibility may depend on the perpetrator satisfying certain conditions, such as disclosing the full truth about the abuses in which s/he is implicated. *Reparations*, which can be material and/or symbolic, repair (some of) the harm suffered by victims and affirm the status of victims as rights bearers and equal members of the political community who deserve better treatment.[12] *Memorials*, in the case of transitional justice, represent

[8] Nir Eisikovits, "Transitional Justice," in Edward N. Zalta (ed.), *The Stanford Encyclopedia of Philosophy* (Fall 2017 edition).

[9] For an overview of ongoing work, see the International Center for Transitional Justice, www.ictj.org.

[10] Priscilla Hayner, *Unspeakable Truths: Transitional Justice and the Challenge of Truth Commissions*, second edition (New York: Routledge, 2010).

[11] Mark Freeman, *Necessary Evils: Amnesties and the Search for Justice* (New York: Cambridge University Press, 2009).

[12] Pablo de Greiff (ed.), *Handbook of Reparations* (New York: Oxford University Press, 2008). For discussion of the relation between reparations and reconciliation, see Daniel Philpott's chapter in this volume.

a standing commitment to acknowledge in an ongoing and permanent manner events that were wrongful and victims who suffered.

There are five widely recognized pillars of transitional justice, which reflect the specific normative goals such processes serve.[13] (1) *Justice* captures the aim of holding perpetrators to account for wrongful conduct; this pillar is frequently linked with criminal trials. (2) *Truth* reflects the right of victims to know what happened to them and who was responsible, and to have these truths acknowledged in a public manner. Truth is important in contexts where the occurrence or extent of wrongdoing is officially denied, and the commission of wrongdoing occurred with impunity. (3) *Reparations* captures the need to prioritize victims and their needs, both moral and material, in the aftermath of wrongdoing. Prioritization of victims matters in contexts where victims are in many cases from marginalized communities and where the expressive purpose of wrongdoing is to deny that victims are rights bearers and equal members of a given political community. (4) *Nonrecurrence* reflects the aspiration to establish conditions that will ensure that atrocities of the past do not occur again in the future. The starting premise of transitional justice is that by dealing with the past, communities can ensure that "Never again" becomes a reality. (5) *Memory*, the recently added fifth pillar, points to the need to acknowledge, not just at the time processes are established but in a more permanent and enduring manner, the atrocities on which transitional justice processes focus.[14]

These five pillars are not independent but are interconnected. Together they seek to achieve the overarching normative aim of what I call in my work "the just pursuit of societal transformation."[15] I define "societal transformation" relationally. The overarching objective of transitional justice processes is to transform how members of a political community interact with one another and with government officials. Such transformation links transitional justice with reconciliation, the process of rebuilding damaged relationships.[16] Relational transformation is necessary to address the underlying conditions that enabled atrocities, thereby making 'nonrecurrence'

[13] United Nations, "Guidance Note of the Secretary General: United Nations Approach to Transitional Justice," March 2010, https://digitallibrary.un.org/record/682111.

[14] United Nations Special Rapporteur on Truth, Justice, Reparations, and Guarantees of Nonrecurrence, "Memorialisation Processes: Report," July 9, 2020, https://documents-dds-ny.un.org/doc/UNDOC/GEN/G20/175/70/PDF/G2017570.pdf?OpenElement.

[15] Colleen Murphy, *The Conceptual Foundations of Transitional Justice* (Cambridge: Cambridge University Press, 2017).

[16] Colleen Murphy, *A Moral Theory of Political Reconciliation* (Cambridge: Cambridge University Press, 2010); Colleen Murphy and Linda Radzik, "Reconciliation" in Edward N. Zalta (ed.), *The Stanford Encyclopedia of Philosophy* (Winter 2020 Edition).

practically possible. Relational transformation is only possible if there is first knowledge and acknowledgment of the truth about the causes of past wrongdoing.

The just pursuit of transformation signals the importance of ensuring processes of transitional justice do not treat immediate subjects (e.g., perpetrators and victims) as mere means for the sake of some broader societal end. Perpetrators and victims who are subjects and/or participants in processes of transitional justice must be treated as befits their status as victims or perpetrators. As such, victims must be acknowledged as having been wronged and as entitled to better treatment. The wrongful actions of perpetrators must be condemned, and responsibility for those actions placed upon perpetrators.

The need to pursue transitional justice occurs, I argue in my work, in contexts that have four features.[17] The first is *normalized wrongdoing*. Transitional justice is not necessary unless human rights violations occurred and is only salient when such violations become widespread. "Normalized" captures the consequence of widespread wrongdoing for communities: wrongdoing is normalized when it becomes a basic fact of life around which members of a community (or members of the specific group targeted by wrongdoing) must orient their conduct. Dissidents, for instance, may be reluctant to speak critically of the government for fear of being disappeared. As this example suggests, wrongdoing of interest to transitional justice characteristically implicates both state and nonstate actors. Wrongdoing is normally political, not simply ordinary criminality, but rather oriented toward the pursuit of certain political goals. Defending a regime, winning a conflict, acquiring land, and silencing dissent are examples of the kinds of political purposes wrongdoing serves. Widespread wrongdoing has also a collective dimension. The killing of many is often possible only when carried out by groups, and the many killed are in numerous cases targeted or vulnerable to wrongdoing because of their race, ethnicity, gender, geographic location, or political ideology.

This first feature of transitional contexts can, but does not necessarily, overlap with the contexts of interest to just war theory. In some cases, normalized wrongdoing occurs in the context of violent conflict and constitutes war crimes or crimes against humanity, violations of the *jus in bello* constraints on conduct in warfare. Normalized wrongdoing occurs in the background of ongoing repression and/or conflict and may generate grounds for humanitarian intervention. However, normalized wrongdoing

[17] Murphy, *Conceptual Foundations*, chapter 1.

does not always constitute war crimes or crimes against humanity, nor does repression always suffice to justify humanitarian intervention.

Many factors can contribute to the normalization of wrongdoing. One important factor is the general impunity for violating human rights that perpetrators enjoy. Impunity can reflect the weakness of legal institutions, where effective recourse for victims of human rights violations is absent. The absence of protection can be a function of corruption in law enforcement, where members of law enforcement prioritize protecting their own over holding to account members who engage in abusive action. It can be a function of policies that sanction or require violations. The "false positives" scandal in Colombia involved members of the Colombian military dressing up civilians they killed as combatants to artificially boost the reported numbers, and so perceived success, of the military during the conflict.[18] It can be a function of corruption within the judiciary. It can be a combination of all these factors. Whatever the differences in the specific factors that lead to impunity, the absence of meaningful accountability for violations of human rights is shared.

A second factor shared by transitional societies is what I call *pervasive structural inequality*. "Structural" references institutions that shape interaction by defining permissible, prohibited, and required conduct as well as the formal and informal sanctions for violating these rules and norms on conduct. Institutions of interest include legal, political, economic, social, cultural, and religious bodies. Institutions refer to social practices of cooperation, which can be more or less formal and which exist as a result of a system of interconnected rule-governed roles.[19]

The "inequality" of interest is two-fold. One dimension is in terms of how institutions, in their explicit structure or as a consequence of their operation, differentially impact what individuals in a community have a genuine opportunity to do or become.[20] The genuine opportunity to own land may be affected by legal rules, which may explicitly prohibit individuals as a matter of ethnicity, race, or gender from owning property. Protection of property owned may vary; the effective ability of women to hold on to property owned against encroachment from family members

[18] Joe Parkin Daniels, "Colombia Tribunal Reveals at Least 6,402 People were Killed by Army to Boost Body Count," *The Guardian*, February 19, 2021, www.theguardian.com/global-development/2021/feb/19/colombia-farc-tribunal-false-positives.

[19] Seamus Miller, "Social Institutions" in Edward N. Zalta (ed.), *The Stanford Encyclopedia of Philosophy* (2014).

[20] This language is drawn from the capability approach pioneered by Martha Nussbaum and Amartya Sen. See Ingrid Robeyns and Morten Fibieger Byskov, "The Capability Approach," in Edward N. Zalta (ed.), *The Stanford Encyclopedia of Philosophy* (Winter 2020 edition).

or other members of the community is weak in many communities.[21] Redlining policies in the United States prohibited Black men and women from owning property in designated areas, with ripple effects for public educational opportunities. Campaigns of racial terror like the Tulsa Race Massacre left property-owning Black individuals vulnerable to destruction with little recourse for protection or compensation from the state.[22]

A second dimension of inequality is in terms of the differential ability of various groups in a political community to shape the institutional rules and norms themselves. Consider the right to vote. Certain groups of individuals may be formally barred from voting in societies where the right to vote exists for some. This was true of Black South Africans during apartheid. Voter suppression tactics can undermine the genuine opportunity to vote, by imposing unreasonable time requirements to cast a vote, paperwork requirements to prove one's identity, and/or geographic impediments by placing voting booths at great distances from where certain communities live. The threat of violence for voting can also serve as a deterrent. In the Sierra Leone civil war from 1998–1999, limb amputation was the consequence for some who attempted to vote.[23]

Such inequality can both reflect and underscore the unequal status of certain marginalized communities within a society. It can reflect and produce erosion of the human rights entitlements that members of such communities should enjoy. Normalized wrongdoing and pervasive structural inequality are often mutually reinforcing. In the background of the structure of rules governing conduct and the normalization of wrongdoing is often a justifying ideology. In cases of what philosopher Tommie Shelby calls ideological racism, commonly held beliefs and judgments "explaining" social reality (e.g., differential educational attainment across racial groups or the enduring racial gap in the United States) are distorting in ways that perpetuate injustice.[24]

The third general factor in transitional societies is what I call *serious existential uncertainty*. Efforts to pursue transitional justice typically occur in the aftermath of the signing of a peace agreement, toppling of a dictatorship, or transition to democratic institutions after a period of apartheid.

[21] Colleen Murphy and Charles Fogelman, "Gender, Structural Inequality, and Just Governance," in Uchendu Chigbu (ed.), *Land Governance and Gender: The Tenure-Gender Nexus in Land Management and Land Policy* (CABI International, forthcoming).

[22] Richard Rothstein, *The Color of Law: A forgotten history of how our government segregated America* (New York: Liveright, 2017).

[23] Sierra Leone Truth and Reconciliation Commission, "Witness to Truth: Final Report of the TRC," 2004, www.sierraleonetrc.org/index.php/view-the-final-report/download-table-of-contents/volume-one.

[24] Tommie Shelby, "Ideology, Racism, and Critical Social Theory," *The Philosophical Forum*, 34, no. 2 (2003), 153–88.

Thus, they are established in moments when a sign of concrete change exists. Yet the period of change is exceptionally fragile. Peace agreements may not signal the end of conflict in the near or long term. In Colombia, the ongoing killing of human rights activists after the signing of the Final Agreement to end conflict remains a pressing issue.[25] In Northern Ireland, the tensions Brexit has placed upon the Belfast/Good Friday Agreement formally ending The Troubles has resulted in a more visible presence of paramilitary forces.[26] Many attempted transitions away from conflict, repression, and injustice fail, underscoring the fragility of gains.

Uncertainty can arise where it has not existed before. The "Big Lie" former President Donald Trump has perpetuated about the 2020 United States Presidential election (specifically, the unsupported claim that there were systematic irregularities in the voting process and that he won the election) led to uncertainty, for the first time in recent memory, as to whether in fact there would be a peaceful transition from one administration to another.[27] The January 6 Capitol insurrection only further underscored the uncertainty. Subjectively, it is very difficult for members of transitional societies and observers to know what narrative to tell about unfolding events. Competing narratives about how well the institutions of the United States performed in the face of the threat of Trump and the Capitol insurrection are one example.

Finally, *fundamental uncertainty about authority* reflects the precarious nature of the standing of the state to deal with rights violations in which it is implicated and the often-disputed authority of certain officials or officials in general to govern given the ongoing but not yet consolidated political shift that is taking place (e.g., transition to democracy).

The Critical Turn

Despite their normative ambitions, both just war theory and transitional justice have become subjects of a critical turn. The critical turn in both cases contests whether each framework is conceptually adequate for dealing with the contemporary form that their subject matters take. Such conceptual deficiencies ground skepticism over the aptness of the normative guidance each provides. There are two conceptual limits to each framework that

[25] Human Rights Watch, "Amicus Brief on Killings of Human Rights Defenders in Colombia," April 20, 2021, www.hrw.org/news/2021/04/20/amicus-brief-killings-human-rights-defenders-colombia.

[26] Michael Hirst, "Northern Ireland Riots: What Is Behind the Violence in Northern Ireland?" *BBC News*, April 14, 2021, www.bbc.com/news/uk-northern-ireland-56664378.

[27] Timothy Snyder, "The American Abyss," *The New York Times*, January 9, 2021, www.nytimes.com/2021/01/09/magazine/trump-coup.html.

I want to highlight. The first limit is in the temporal representation of the phenomena of interest.

In the case of just war theory, the existence of what have come to be called "forever" or "endless" wars challenges the implicit theoretical assumption that war is a temporally discrete occurrence with a definite beginning and clear end.[28]

In the case of transitional justice, the worry is articulated in terms of an artificial line the "transitional" nature of justice draws between a "pretransition" past and a "post-transition" period. The idea of "transition" suggests that transitions are radically disruptive moments. However, in reality, transitions are often anything but radically disruptive, and often a meaningful transition either has not or will not occur.

Consider paradigm cases where there is an event that putatively ends conflict or repression. This event could be in the form of a peace treaty that formally declares an end to violence or the toppling of a repressive regime. Even when such momentous events occur, dynamics of conflict and repression from the past continue. In the case of conflict, violence may decrease, but not be eliminated. In contexts like Colombia, formal peace agreements co-exist with the assassination of human rights defenders. In cases of repression, the conditions that enabled repression remain, even with a dictator gone. The end of Hosni Mubarak's reign in Egypt did not bring about the end of torture for dissidents in detention.

Further complicating the attempt to individuate temporally discrete "pre-" transition and "post-" transition periods is the fact that transitional justice processes are now used in contexts where no transition to speak of has taken place. Consider Syria, which at the time of this writing continues a decade-long civil conflict between the government of Bashar al-Assad and his opponents. This conflict has led to the displacement of millions of Syrians and the deaths of hundreds of thousands more. Torture is endemic, despite the fact that transitional justice as a concept has been invoked, and efforts at accountability have been pursued.[29]

[28] Council on Foreign Relations, "The Iraq War: 2003–2011," www.cfr.org/timeline/iraq-war; Council on Foreign Relations, "The U.S. War in Afghanistan: 1999–2021," www.cfr.org/timeline/us-war-afghanistan; David Zucchino, "The War in Afghanistan: How It Started and How It Is Ending," *The New York Times*, July 31, 2021, www.nytimes.com/article/afghanistan-war-us.html.

[29] Anwar el Bunni, "Breaking new ground: Transitional justice in Syria," Brookings Institute, November 24, 2020, www.brookings.edu/opinions/breaking-new-ground-transitional-justice-in-syria/; International Center for Transitional Justice, "Syria," www.ictj.org/our-work/regions-and-countries/syria; Noha Aboueldahab, "Writing Atrocities: Syrian Civil Society and Transitional Justice," Brookings Doha Center, Analysis Paper Number 21 (May 2018), www.brookings.edu/research/writing-atrocities-syrian-civil-society-and-transitional-justice/.

Transitional justice has also been pursued in contexts like Canada, where no transition, as traditionally understood, has occurred. There is no peace agreement to mark the end of conflict or a change of the structure of government. Instead, the Truth and Reconciliation Commission of Canada from 2007–2015 "provided those directly or indirectly affected by the legacy of the Indian Residential Schools system with an opportunity to share their stories and experiences."[30] These cases underscore the continuity, rather than disruption, that characterizes the situations in which the pursuit of transitional justice occurs.

The worry with the temporal representation is not only that it leads to a paradigm that is descriptively inadequate, but also that it produces problematic normative prescriptions. The descriptive picture of a radical break may suggest that this break is achieved where it does not yet exist. Consider law enforcement and the military, which are frequently the subject of reform efforts. Purges may seem to facilitate a radical break with prior law enforcement. However, the experience from the Balkans, for instance, shows that purges may create more problems than they resolve, as unemployed members of law enforcement simply find alternate ways to use their skills to secure an income. Fired police became mercenaries in the Balkan Wars.[31] While purging the judiciary may not produce that problem, it can lead to a vacuum of expertise needed for a judicial system to function. The policy of de-Baathification left an Iraqi judiciary with extremely old and extremely young judges, judges with little judicial experience, on the one hand, and experienced judges with little knowledge of law and its operation during the 25-year reign of Saddam Hussein.[32] Most transitional societies avoid complete purges, choosing to vet security forces for ties to extremist groups through programs of lustration or firing some but not all judges.[33] Police forces may be renamed but contain prior members. While it is understandably a source of consternation to some victims to see the same faces in institutional roles that were present prior to a transition, such continuity may

30 Truth and Reconciliation Commission of Canada, www.rcaanc-cirnac.gc.ca/eng/1450124405592/1529106060525. The final report is available at: https://nctr.ca/records/reports/.

31 Genevieve Bates et al., "What is the effect of Personnel Transitional Justice on Crime?" www.monikanalepa.com/uploads/6/6/3/1/66318923/draft_jan8_20.pdf.

32 Miranda Sissons and Abdulrazzaq Al-Saiedi, "Iraq's de-Baathification still haunts the country," *Al Jazeera*, March 12, 2013, www.aljazeera.com/opinions/2013/3/12/iraqs-de-baathification-still-haunts-the-country/; Miranda Sissons and Abdulrazzaq Al-Saiedi, "A Bitter Legacy: Lessons of De-Baathification in Iraq," *International Center for Transitional Justice Report*, March 4, 2013, www.ictj.org/publication/bitter-legacy-lessons-de-baathification-iraq.

33 Monika Nalepa, *Skeletons in the Closet: Transitional Justice in Post-Communist Europe* (New York: Cambridge University Press, 2010).

be necessary, all things considered. Efforts at transformation that are maximally disruptive may not in fact be conducive to transformation.

A second line of critique concerns the omission of power dynamics in the theory of just war and of transitional justice. Yet power imbalances profoundly shape what the pursuit of justice looks like in the aftermath of war and in transitional justice. The process of pursuing justice for wrongdoing is not shaped by equally situated actors operating in a context of mutual respect and against a background of justice. Instead, accountability is imposed by victors on the vanquished. Even where there are no vanquished, the unjust but powerful, or the powerful outsider, dictate whether and how "justice" gets pursued. An underlying source of the problems in these cases is that the wrong actors are deciding what justice looks like. Asymmetrically situated parties result in powerful actors being able to shift the gaze of justice away from their own actions.

Just war theory and transitional justice both reject realism; both are predicated on the assumption that morality still applies even in highly imperfect conditions. However, a rejection of realism is compatible with a recognition of the normative significance of power dynamics. By assuming away such power dynamics, accounts are not only descriptively inadequate, but also conceptually erase normative issues. The normative worry is that power imbalances compound injustice rather than correct it through processes of accountability.

Consider which wrongs become the subject of *jus post bellum* and/or transitional justice processes. Many processes focus on specific human rights violations that occurred over a delimited period of time, and which were committed by identifiable perpetrators. For example, the South African Truth and Reconciliation Commission (TRC) focused on killing, torture, abduction, and severe ill-treatment committed by agents of the South African state or by members of armed anti-apartheid organizations from 1960–1994.[34] While undeniably wrong, the choice of extraordinary wrongdoing committed by both sides distorted the central wrong of *apartheid* itself; apartheid is one of the crimes against humanity. The central wrong of apartheid was the set of legally institutionalized rules and norms that segregated land and in turn justified the forcible removal of millions of Black South Africans from their homes, set up educational systems explicitly aimed at unequal education along racial lines, and prohibited meaningful political participation on the part of Black South Africans.

[34] South African Truth and Reconciliation Commission, "Final Report" (1998), www.justice.gov.za/trc/.

Put differently, as critics of the South African TRC have noted, the subject matter of the TRC effectively placed the structural injustice of apartheid as the background context for transitional justice, instead of as the subject of transitional justice processes.[35] Structural injustice is a function of the terms for interaction and those that structure interaction.[36] Not only the rules and norms of institutions, but the underlying ideology that provides the justification for such rules and norms can contribute to structural injustice. As Catherine Lu illustrates with the ideology of "civilization," a racialized hierarchy shaped who was judged "civilized" and who "uncivilized," where the civilized were equated with the "white and European." This ideology provided resources to diminish recognition of the rights and claims of non-white people. By failing to address apartheid, measures of transitional justice left intact the basic infrastructure of material, social, and economic inequality and the material consequences of decades of apartheid policy.

Similarly, according to critics, historical wrongdoing is also unjustifiably overlooked by the theory and practice of transitional justice. Historical wrongdoing does not have identifiable primary victims who may still be alive, nor perpetrators who could be held accountable via a process like a criminal trial. Yet historical wrongdoing in many contexts continues to shape interaction and reproduce patterns of inequality – as within communities there exist both beneficiaries of past wrongdoing and bearers of past wrongdoing's harmful consequences – and so is implicated in it.[37]

Power also shapes *who* gets held to account by processes. In the context of postwar justice, critics of the Nuremberg Trials point to the fact that Allied judges exclusively tried non- Allied officers for crimes; in the International Military Tribunal for the Far East in Tokyo, Allied judges tried only Japanese war criminals.[38] The result was that no historical record of Allied crimes was produced.

Finally, powerful external actors, especially development agencies pursuing their own goals in postwar or transitional justice circumstances, alter the agenda of justice.[39] On this last point, critiques in the case of

[35] Mahmood Mamdani, "Amnesty or Impunity? A Preliminary Critique of the Report of the Truth and Reconciliation Commission of South Africa," *Diacritics*, 32, no. 3/4 (2002), 33–59.

[36] On the idea of structural wrongdoing see Catherine Lu, *Justice and Reconciliation in World Politics* (New York: Cambridge University Press, 2017); Iris Marion Young, *Justice and the Politics of Difference*, rev. ed. (Princeton, NJ: Princeton University Press, 2011).

[37] Michael Rothberg, *The Implicated Subject: Beyond Victims and Perpetrators* (Stanford, CA: Stanford University Press, 2019).

[38] Padraig McAuliffe, "Transitional Justice and the Rule of Law: The Perfect Couple or Awkward Bedfellows?" *Hague Journal on the Rule of Law*, 2 (2010), 127–54.

[39] Pablo De Greiff and Roger Duthie (eds.), *Transitional Justice and Development: Making Connections* (New York: Social Science Research Council, 2009).

transitional justice discuss the ways in which the agenda and priorities of development agencies, foreign governments, and international transitional justice NGOs dictate (or have a disproportionate influence over) which wrongs get addressed by transitional justice and in what way.[40] This undermines the agency of local actors. This erasure of possibilities for the exercise of agency is problematic in itself. The absence of agency of local actors in shaping transitional justice also results in poorer decision-making. There can be a disconnect between the priorities of international and domestic actors. In some cases, there is not just difference, but also incompatibility between priorities. International actors may furthermore act on the basis of insufficient knowledge of the local context in which they are operating.

A cumulative consequence of the conceptual limitations discussed above is the articulation of two normative complaints that theories of just war theory and transitional justice must provide resources to address. The first is a complaint that victims have been sold short from the perspective of justice.[41] Victims of crimes against humanity, war crimes, torture, or displacement fail to receive the justice that is due, because, partly or wholly, of the distortions of the pursuit of justice to which the power asymmetries discussed above give rise. "Victims" here is understood broadly to include those explicitly acknowledged in processes of transitional justice or *jus post bellum* as well as those whose victimization goes unacknowledged.

The second is a complaint that rather than further justice, the unjust status quo is entrenched by processes of transitional justice or postwar accountability.[42] As the Taliban now control Afghanistan, the question of what, if anything, American occupation for the past two decades has achieved has become especially acute.[43] The same question about post-apartheid South Africa is implicit in ongoing protests following Zuma's arrest. As Dustin

40 Sirkku Hellsten, "Transitional Justice and Aid," UNU/WIDER Working Paper 2012/06, UNU/Helsinki, www.wider.unu.edu/sites/default/files/wp2012-006.pdf.

41 On the complaint that victims are sold short, see Luke Moffett, "Reparations in Transitional Justice: Justice or Political Compromise?" Queen's University Belfast School of Law: Research Paper 2019-07 (2017), 1–17, https://pureadmin.qub.ac.uk/ws/portalfiles/portal/120688204/Reparations_in_Transitional_Justice.pdf; Warren Buford and Hugo van der Merwe, "Reparations in Southern Africa," *Cahiers d'etudes africaines*, 44, no. 1/2 (2004), 2–53. For an initial articulation of the challenges I discuss here and the criteria for principled compromise, see Colleen Murphy, "On Principled Compromise: When does a process of transitional justice qualify as just?" *Proceedings of the Aristotelian Society*, 120 (2020), 47–70.

42 Dustin Sharp, "What Would Satisfy Us? Taking Stock of Critical Approaches to Transitional Justice," *International Journal of Transitional Justice*, 13, no. 3 (2019), 570–89.

43 Julian E. Barnes, Thomas Gibbons-Neff, and Eric Schmitt, "Officials Try to Sway Biden Using Intelligence on Potential for Taliban Takeover of Afghanistan," *The New York Times*, April 23, 2021, www.nytimes.com/2021/03/26/us/politics/biden-afghanistan-intelligence.html?searchResultPosition=1.

Sharp writes, "while transformation may be a useful conceptual prism for thinking about the limitations of the mainstream goals and modalities of the field, taken literally it is an improbable outcome in most transitional justice scenarios, and perhaps especially in fragile postconflict states..."[44]

In the face of daunting power inequalities that shape justice and the limited impact that military intervention and transitional justice processes have had, what should our reaction be? The challenge, I suggest in the next section, is to locate the bar for minimal normative success.

Minimal Thresholds for Normative Success

In addition to articulating a standard of full or perfect justice, the floor for just intervention or processes of transitional justice must be articulated. Such a floor will allow us to distinguish between processes of transitional justice or postwar justice that sell victims short by failing to achieve any measure of justice at all, on the one hand, and processes that achieve justice, though not completely, on the other. To respond to complaints of victims that no justice has been rendered, it is necessary first to specify what would be necessary to achieve justice to any degree at all. This worry about the absence of meaningful justice being achieved is especially intense in the severely constrained circumstances in which postwar reconstruction efforts and processes of transitional justice normally occur. Background impunity and existing unequal power dynamics further reinforce the concern of victims that meaningful accountability or repair of harm will not occur. A way of demonstrating whether and how moral values have been considered is necessary, as is a way of ensuring the fact that the values being taken into account are recognizable by victims. Similar considerations hold with respect to broader societal transformation. To rebut a complaint that after transitional justice or postwar efforts the context is the same or perhaps even worse, it is necessary to articulate what it would mean to change a context for the better, even to a minimal degree. And this standard for change must be made recognizable to broader members of the community in which transitional justice or postwar reconstruction efforts are taking place.

To explain the idea of a normative floor, I draw on a discussion by legal philosopher Lon Fuller with respect to the normative value of the rule of law.[45] What matters for my purpose is the two-level structure of evaluation he proposes. In *The Morality of Law*, Fuller distinguishes the

[44] Sharp, "What Would Satisfy Us?" 102.

[45] Lon Luvois Fuller, *The Morality of Law*, rev. ed. (New Haven, CT: Yale University Press, 1969).

morality of duty from the morality of aspiration. The morality of duty sets a floor of obligations that must be met for moral behavior. Insofar as the obligations laid out in the morality of duty have not been satisfied, wrongdoing has taken place. In the morality of aspiration, the goal of action is excellence, not merely the satisfaction of minimal standards for moral behavior to have occurred. Failing to meet standards of excellence involves no wrongdoing, but success is measured by proximity to the aspiration.

Fuller uses the distinction between the morality of duty and morality of aspiration to illustrate two different perspectives from which we can evaluate the rule of law. For Fuller, law is the governance of conduct on the basis of rules. The rule of law, according to Fuller, articulates the conditions for such governance to be possible. We (legal subjects and officials) can only govern our conduct according to rules if in fact these rules can figure in our practical deliberation. To so figure, rules must be known (or promulgated), prospective (not retroactive), clear, constant, and not demand the impossible. For rules, and not the whims of officials who govern, there must also be congruence between what declared rules say and how officials act.

From the perspective of the morality of duty, we can identify a minimum threshold level for each of the eight requirements that the rule of law must satisfy for law to exist in any meaningful sense. As Fuller writes, "A total failure in any one of these eight directions does not simply result in a bad system of law; it results in something that is not properly called a legal system at all, except perhaps in the Pickwickian sense in which a void contract can still be said to be one kind of contract."[46] It is the duty of officials in a system that claims to govern by law to meet this minimum threshold level. They are guilty of failure with respect to the obligations insofar as they fail to meet this standard.

Many systems of governance are legal systems in which rules satisfy the minimum threshold of each of the eight criteria. However, evaluations of the rule of law from the perspective of the morality of aspiration consider how well such laws govern. For this examination, it matters how robustly the criteria of the rule of law are met. Systems of law vary in the degree of promulgation that exists, as well as the extent of prospectivity one finds. The congruence between declared rules and official action can vary. How much the rule of law obtains requires consideration of legal systems as a whole, and the ways in which the systems of

[46] Ibid., p. 39.

rules effectively govern conduct. Judgment is ineliminable in assessing a system of law; there may be tensions between constancy and clarity. Some violations of the rule of law may not be inimical to law's purpose. Fuller illustrates this latter point using the example of a printing press that broke down, but was the only printing press from which official marriage certificates could be issued. During the period in which it was getting fixed, marriages occurred, and retroactive licenses were issued once the press was fixed. Such retroactivity did not undermine the rule of law.

Fuller's analysis offers a model for thinking about the realization of justice both postwar and in transitional justice. It suggests we first think of the conditions that must be satisfied at a threshold level for the label of justice to be applicable. The threshold kind of reconstruction that must occur postwar, the degree of accountability that must be achieved, the minimum justice victims can demand: These are all specific variables that will shape whether we can talk about postwar "justice" and transitional "justice." Failure to meet these duties will result in interventions that merely produce further discord and processes for dealing with wrongdoing that offer victims blood money or mere words instead of repair.

I end this section by considering an objection to the proposal for specifying minimal thresholds. The objection is that this incentivizes actors involved in transitional justice or postwar efforts to aim for the floor, rather than aspiring for the ceiling. This incentive effect is likely to be especially pronounced given the inequality in power noted in the previous section. Powerful actors wishing to avoid accountability for wrongdoing or wishing to skirt more demanding obligations for postwar reconstruction, for example, can use the minimum moral thresholds account to obstruct more robust efforts of justice.

In response, I grant the underlying worry contained in the objection. It may well be that the wrong moral incentives are created for powerful actors if moral floors are specified. However, the force of the worry is limited for two important reasons, both of which show that the worry is not sufficient to undermine the need to articulate minimum moral thresholds.

The first reason the worry is insufficient to undermine the proposal is that the absence of such floors does not result in the robust pursuit of transitional justice or postwar justice. This was the point of the previous section. Transitional justice processes, considered discretely, have limited impact on broader societal transformation, and the worry from occupation efforts in contexts like Iraq and Afghanistan is that robust, aspirational commitments to democratic transformation fail to be realized. So my proposal is not interfering with success. Instead, it is predicated on a recognition of the current reality.

The second reason the worry is overstated is that it rests upon a limited understanding of the purpose of moral thresholds. After we have set the floor for justice to be present in any measure at all and answered this basic threshold question as to whether the floor is sufficiently likely to be met, we can go on to ask whether enough justice has been achieved in any given postwar or transitional context. Whether "enough" justice will or has been achieved is a complicated issue. Answering that question requires more than simply pointing to a minimum floor that has been surpassed. At the same time, "enough" justice is not the same as "full" or complete justice. In the highly constrained circumstances in which just war theory and transitional justice operate, full justice remains an unattainable aspiration. "Enough" justice is a function of achieving what could reasonably be asked or demanded in the circumstances. To identify what this means in any particular case will require, I have suggested elsewhere, formulating standards for principled compromises among the competing and at times conflicting normative demands that arise in war and in transitional justice under severely constrained circumstances.[47] In the case of transitional justice, for example, "enough" is a function of balancing demands for accountability for perpetrators, reparations for victims, reform of institutions, and the nontransitional justice normative demands that societies emerging from periods of conflict and/or repression must also satisfy.

Conclusion

In this chapter, I have examined two structurally analogous descriptive, conceptual limitations that current theorizing about postwar justice from the perspective of just war theory and transitional justice share. Both conceptualize, often implicitly, divisions into discrete periods of "war" and "postwar," and "pretransition," "transition," and "post-transition." Both also fail to weigh the profoundly unequal power dynamics that shape decision-making and the pursuit of justice after war and in transitional justice. These descriptive limitations have normative consequences. They bracket important questions of justice that the practices of postwar justice

[47] On this point, see Murphy, "On Principled Compromise."

and transitional justice raise, for example, the choices of wrongs to address, the (in)justice victims in fact have received and do receive, and the practical possibility of postwar and transitional justice. To engage and provide guidance for dealing with the normative complaints present in contexts of postwar reconstruction and transitional justice, I have argued, it is necessary to specify minimal moral thresholds that need to be met for a value like justice to have been promoted at all, let alone to any substantial degree.

CHAPTER 6

Ending Endless Wars

Alex J. Bellamy

Over the past few decades, it has become increasingly difficult for major powers to translate battlefield victories into decisive and favorable political outcomes. As a result, U.S. military engagements in the Middle East, Russian engagements in its "Near Abroad" and in Syria, French engagements in sub-Saharan Africa, and the African Union's war in Somalia have turned into protracted missions with little prospect of decisive victory. This chapter examines the phenomenon of the so-called "endless war," asking why it has become so difficult for major powers and others to bring wars to an end and what can be done about it. I argue that the problem is a global one rooted in the changing nature, purposes, and attitudes of war. As wars become less about resolving disputes between states and more about the internal composition of states, and as those contests become ever more internationalized, the capacity of actors to sustain war over long periods has increased while incentives to pursue peace have declined. Addressing the problem requires more than just the recalibration of U.S. and Western foreign policies. Instead, a more thorough rethinking of strategic issues and global institutions is required. To make the argument, I proceed in three parts. The first examines the "endless war" thesis which grounds the problem in U.S. liberal hegemony, finding that it does not correspond well with the data. The second offers a brief explanation of some of the factors that extend a war's duration and inhibit peace. The third and concluding section offers some thoughts on how these issues might be addressed.

Understanding Endless Wars

The problem of the "endless war" is commonly depicted as a largely, if not exclusively, American creation of the post-Cold War era – though at least one important study charts its rise to the beginnings of the Second

World War.[1] The phenomenon has inspired a raft of books and articles, and even a new Washington-based think-tank – the "Quincy Institute for Responsible Statecraft," inspired by John Quincy Adams's famous aphorism that the U.S. "goes not abroad in search of monsters to destroy" and a guiding sense of anxiety that this is precisely what the U.S. *has* done since the end of the Cold War. Exponents proffer two principal theories to explain endless wars, both grounded in the idea of America's liberal overreach. For John Mearsheimer, the problem rests in the particular brand of liberalism adopted by the U.S. and its allies after the collapse of the Soviet Union.[2] For Stephen Walt, the problem lies with the Washington foreign policy establishment's groupthink.[3] But although one is more concerned with how liberalism's flawed theories have encouraged flawed foreign policies and the other with how Washington's liberal foreign policy elite consistently reproduces bad policy, the similarities between them are striking and may be said to constitute an "endless wars thesis."

The thesis holds that successive U.S. administrations squandered opportunities presented by the end of the Cold War by pursuing liberal hegemony – understood as "an ambitious strategy in which a state aims to turn as many countries as possible into liberal democracies"[4] – as a grand strategy and ignoring calls for restraint born of realism. This aggressive pursuit of liberal hegemony drew America and its allies into the endless wars of the 2000s, imposing a heavy cost on the West while also weakening it by provoking local opposition in the form of Islamist insurgencies and great power pushback in the form of more aggressive Chinese and Russian postures. The remedy would be a "return" to realist foreign policy characterized by "restraint" for Mearsheimer,[5] and "offshore balancing" for Walt.[6]

At the root of the problem, according to this thesis, is the fact that while liberalism presents itself as universal, it is anything but – a point made by E.H. Carr when he identified the purported universalism of the League of Nations as merely the preferences of the powerful masquerading as

[1] Stephen Wertheim, *Tomorrow, the World: The Birth of US Global Supremacy* (Boston, MA: Harvard University Press, 2020).

[2] John J. Mearsheimer, *The Great Delusion: Liberal Dreams and International Realities* (New Haven, CT: Yale University Press, 2018).

[3] Stephen M. Walt, *The Hell of Good Intentions: America's Foreign Policy Elite and the Decline of US Primacy* (New York: Farrar, Straus and Giroux, 2018).

[4] Mearsheimer, *The Great Delusion*, p. 1.

[5] Ibid., p. 217.

[6] Walt, *The Hell of Good Intentions*, p. 255ff.

moral truth and replicated (e.g. Cunliffe)[7] to explain our current predicament.[8] Yet liberalism's mask of universality does not put its values beyond political or moral contestation and even the greatest of powers cannot impose its will on everyone else. As such, when a global hegemon embarks on a grand strategy that involves using force to create a liberal world in its own image, this will inevitably provoke conflict: the endless wars we see today.

Against this backdrop, the new realist orthodoxy insists that since the end of the Cold War the U.S. and its allies have pursued an aggressive strategy of regime change aimed at spreading liberal democracy in the name of a rules-based liberal order. The Soviet Union's demise encouraged the U.S. to think of its own values as universals and to replace foreign policy guided by the pragmatic accommodation of difference with one focused on the extension of liberal order.[9] Proponents argue that successive administrations – those of G.H.W. Bush, Clinton, G.W. Bush, Obama, and Trump – employed U.S. power not primarily to protect American interests but to expand liberalism, specifically by promoting democracy and human rights, multilateral institutions (for Mearsheimer though not Walt), and market economies and free trade. This was always doomed to fail, they argue, since nationalism – a group's commitment to its own values and interests – and power politics always trump liberal universalism. According to this view, the liberal hegemon's attempt to impose liberalism by force was always going to provoke a backlash. And it is this backlash that we see in the phenomenon of the endless war. These wars are endless since they pit the immense power and obduracy of a hegemon convinced its values are universal against a group's deep-seated commitment to its own, quite different, values, identities, and interests.

The thesis artfully explains the apparent rise of the endless war and crisis in the international order.[10] Yet it only works if endless wars are indeed a primarily American problem and if we can see a pattern of endless war emanating from the aggressive pursuit by the U.S. and its allies of liberal democracy and human rights through forcible regime change. And this

7 Philip Cunliffe, *The New Twenty Years' Crisis: A Critique of International Relations 1999–2019* (Montreal: McGill-Queen's University Press, 2020).

8 Edward Hallett Carr, *The Twenty Years' Crisis 1919–1939: An Introduction to the Study of International Relations* (London: Palgrave).

9 Patrick Porter, *The False Promise of Liberal Order: Nostalgia, Delusion and the Rise of Trump* (Cambridge: Polity, 2020).

10 A fuller explanation of the latter is offered by Alexander Cooley and Dan Nexon, *Exit from Hegemony: The Unraveling of the American Global Order* (New York: Oxford University Press, 2020).

is where we reach a problem, because in several important respects, the explanations do not correspond with the data.[11]

The first problem is that endless wars are not a uniquely or even primarily American or Western problem. As the Appendix shows, of the twenty-four ongoing armed conflicts involving more than 500 deaths in 2020, fifteen (65%) had nothing whatsoever to do with the U.S. and its allies. Of these, fourteen were civil wars and one was the interstate conflict between Armenia and Azerbaijan. One cannot therefore draw a seamless line, as the realist theory tends to do, between recent American war experience and broader questions about the nature and direction of international order, for the latter comprises only one part of the former. Many of these wars, though, are not easily categorized as "endless." To calculate that we need to look only at those wars whose durations exceed the norm. On average, since 1945, civil wars of all types have tended to last approximately ten years, an average that has increased steadily with time and generates an expectation that, all other things being equal, a civil war is likely to last between seven and twelve years.[12] Looking only at wars ongoing at the end of 2020 that extend beyond ten years reduces the list to thirteen (Table 6.1). Of these, the majority (eight, or 61%) are civil wars with little or no connection to the U.S. and the West, some of them – such as the wars in Myanmar, Colombia, Turkey, and the DRC – are wars that have endured for decades. Four (30% of the total) ongoing "endless" wars are in some way connected to the War on Terror. These are evenly divided between the two main wars initiated by the U.S. and its allies after 9/11 (the wars in Afghanistan and Iraq) and two transnational civil wars involving Islamist actors (Somalia and Mali) that were not initiated by the West. Though the U.S. and its allies became involved in these wars, mainly through stand-off strikes, special forces raids, and support to host governments (the latter being the very opposite of "regime change") – involvement that may well have extended the war's duration by preventing Islamist victories without (yet) enabling government victories – each of these wars predated U.S./Western intervention and was thus not caused by it. The remaining "endless war" is the territorial dispute between Armenia and Azerbaijan. The endless war phenomenon is thus not unique to the U.S./West, nor is it obvious that the U.S. is its primary cause. These wars are not uniformly

[11] See Robert Jervis, "Liberalism, the Blob, and American Foreign Policy: Evidence and Methodology," *Security Studies*, 29, no. 3 (2020), 434–56.

[12] James Fearon, "Why Do Some Civil Wars Last So Much Longer Than Others?" *Journal of Peace Research*, 41, no. 3 (2004), 275–301; James Fearon and David D. Laitin, "Ethnicity, Insurgency, and Civil War," *American Political Science Review*, 97, no. 1 (2003), 75–90.

Table 6.1 *Today's Endless Wars (ongoing wars extending beyond the average civil war duration of 10 years)*

War Type	No.	Examples
Civil wars	8	Nigeria: Boko Haram vs. Govt. Nigeria Mexico: Govt. Mexico (+ allies) vs. drug cartels Sudan: Govt. Sudan vs. JEM, SLA, SRF + others Myanmar: Govt. Myanmar vs. Karen, Shan, Kachin, Rohingya, Rakhine Nigeria: various communal conflicts DRC: Lendu + allies vs. Hema + allies (inc. Govt. DRC), M26/7 vs. Govt. DRC + allies vs. FDLR vs. NDC-R, Govt. DRC vs. ADF Turkey: Govt. Turkey vs. PKK Colombia: Govt. Colombia vs. FARC
Civil/transnational wars with War on Terror component (U.S./allies)	2	Maghreb: AQIM v Govts. of Algeria, Tunisia, Mali, Burkina Faso, Chad, Mauritania, etc. + allies (inc. U.N., U.S.) Somalia: Transitional govt. + allies (inc. Kenya, U.S.) vs al-Shabaab, ISIS + allies
War on Terror wars	2	Iraq: U.S. + allies (inc. Govt. Iraq) vs. various insurgents Afghanistan: Taliban/AQ vs. U.S. + allies
International wars	1	Azerbaijan: Azerbaijan vs. Armenia
Humanitarian/human rights wars	0	

struggles between liberalism and its alternatives. Instead, U.S. experiences of protracted war in Afghanistan and Iraq seem to mirror broader trends in the duration of wars in which the U.S. is not involved.

Agency too is more complex than the realist theory allows. It is not just that other powers besides the U.S. have their hands in shaping multiple ongoing wars, but that most if not all of this involvement cannot be easily categorized as a response to U.S. hegemonic intervention: Russia (in Syria, Libya, and Armenia-Azerbaijan, as well as Ukraine and Georgia which did not make the list owing to the 500 deaths in 2020 cut-off), Turkey (at home, Kurdistan, Syria, Libya, and Armenia-Azerbaijan), and Iran (in Iraq, Syria, Yemen, and the not qualifying Lebanon, Palestine, and Baluchistan). International and regional organizations are also significant actors in several conflicts, the United Nations in the Democratic Republic of the Congo, South Sudan, and Mali, and the African Union leading ground operations against Islamists in support of the government in Somalia, having earlier also played a role in Darfur. In almost all these situations, the intervention of regional powers was driven by primarily regional considerations in situations where the positioning and contribution of the U.S. was largely a background consideration. Viewing the world through a distinctly American lens, the endless wars thesis exaggerates the West's influence on global conflict trends and simultaneously diminishes the agency of others – a long-standing characteristic of international relations theory.[13] The elongation of armed conflict is a global trend with sources that go beyond U.S. behavior.

A second area where the theory of endless war fails to correspond with the data is the U.S. experience itself. The U.S. did not, in fact, become more interventionist after the Cold War, nor was the interventionism it did embark on directed primarily toward spreading liberal hegemony. The theory, recall, hinges on the idea that after the end of the Cold War, the U.S. adopted an ambitious grand strategy of using force to make the world safe for liberal democracy. If that were true, we would expect to see three things: first, an increase in the number of military interventions initiated by the U.S.; second, a shift in the nature of those interventions away from national security missions and toward humanitarian and democracy-promotion missions; and third, some sort of correspondence between this latter type of missions and conflict duration, that is that America's "endless wars" were in fact wars inspired by its liberalism-extending grand strategy. What we actually see, though, is none of those things.

On the first issue – war initiation – it is not clear that America's rise to primacy after the collapse of the Soviet bloc was accompanied by a sharp increase in U.S. global military intervention. Indeed, there was no obvious increase in the number of new wars initiated by the U.S. associated with the end of the Cold War (Table 6.2). Overall, the rate of intervention in the 1990s remained roughly where it had been in the 1980s before declining somewhat in the two decades after that. If the number of new interventions is a measure of interventionism, the U.S. was much less interventionist in the 2010s than it was in both the 1960s and 1980s. The story of the 2000s is one in which U.S. priorities shifted from a brief focus on humanitarian/human rights operations in the 1990s to a focus on

[13] Amitav Acharya, *The End of American World Order* (Cambridge: Polity Press, 2018); Amitav Acharya and Barry Buzan, *The Making of Global International Relations: Origins and Evolution of IR at its Centenary* (Cambridge: Cambridge University Press, 2019).

Table 6.2 *U.S. War Initiations and Duration by Decade*

Decade	New wars	Average duration (years rounded)
1900s	1	1
1910s	9	7.22
1920s	1	1
1930s	0	-
1940s	2	4.5
1950s	4	2.25
1960s	8	5.63
1970s	1	1
1980s	8	1.38
1990s	7	2.14
2000s	6	11.33
2010s	5	3.68

the War on Terror after 9/11. The increase of War on Terror operations accompanied a sharp decline in the frequency of humanitarian/human rights operations in a context where, overall, the frequency of intervention remained at a level broadly consistent with that of the Cold War era.

What we see are patterns of continuity as well as change traversing the period at the end of the Cold War. For example, the roots of the militarization of the U.S. relationship with the Middle East can be traced to the Iranian revolution which preceded the end of the Cold War by a decade. U.S. policy then seems not to have been driven primarily by an urge to spread liberal hegemony but rather by the logics of realist canons such as securing U.S. access to oil at reasonable prices, protecting political allies, and denying the region to communist and Islamist competitors. Offensive realism for sure, but realism nonetheless. "At no time," Andrew Bacevich tells us, "did the promotion of democracy and human rights figure in Washington's Iranian agenda,"[14] an agenda that gave rise to Cold War era military interventions in Iran, Egypt, Lebanon, and the Persian Gulf, and set in train a network of interests, alliance commitments, and entanglements that established the contexts for later interventions. Of America's many interventions in the Middle East, the promotion of democracy and human rights was foregrounded only in the 2003 invasion of Iraq, and even then, the U.S. came to favor order through elite leadership over rule through the ballot box. The first Gulf War was about order not democracy,

[14] Andrew J. Bacevich, *America's War for the Greater Middle East* (New York: Random House, 2016), p. 13.

and the war in Afghanistan was about defeating al-Qaeda. With the possible exception only of Iraq, U.S. interventionism in the Middle East only paid "lip service" to democracy and human rights.[15]

It may be contended, of course, that the number of new interventions in a given decade does not tell us much about interventionism because it gives no consideration to scale. For that, we might look at the overall number of troops deployed overseas or at the number of military casualties. On the first point, the end of the Cold War was associated with a significant reduction in the deployment of U.S. personnel abroad. That trend was reversed to some extent by the wars in Afghanistan and Iraq, but even these operations only brought total U.S. overseas deployments to a level slightly beneath the level immediately preceding the Gulf War.[16] The annual number of military casualties tells a similar story. The U.S. lost 377 military casualties to hostile action in the 1980s compared to 208 in the 1990s, nearly three-quarters of which (147) resulted from the Gulf War. Casualties more than doubled with the start of Operation Iraqi Freedom in 2003 and the intensification in Afghanistan after 2004.[17] This data intuitively corresponds with what we know about America's wars since 1945. Certainly, the wars in Iraq and Afghanistan were large and costly, but so too were the wars in Indochina and Korea in the 1960s and 1970s. America's war in Indochina matches its war in Afghanistan in terms of duration too. My point here is that in terms of the frequency, scale, and cost of American military interventions abroad, the end of the Cold War did not see the sorts of increases the realist thesis would lead us to expect. There was in fact no peak of U.S. interventionism associated with the end of the Cold War. What is more, although new conflicts were initiated as part of the War on Terror after 9/11, even this did not represent a sharp change from past practice. The principal thing that changed after 2001 was the duration of military campaigns. This was almost entirely a function of the War on Terror. Indeed, America's "endless war" is the War on Terror – nothing more, and nothing less.

This point is borne out by considering the principal context and justification for U.S. wars in the post-Cold War era (Table 6.3). If the realist

[15] Ibid., p. 322.

[16] Tim Kane, *Global US Troop Deployments: 1950–2005* (Washington, DC: Heritage Institute, 2006), p. 6.

[17] Congressional Research Service, *American War and Military Operations Casualties: Lists and Statistics*, July 29, 2020, https://crsreports.congress.gov/product/pdf/RL/RL32492/33, p. 6.

Table 6.3 *U.S. Wars 1990–2020 by Type and Duration*

Type	Number	Average Duration
Collective self-defense (U.N.-mandated)	1	2
Humanitarian and human rights	U.N.: 5 Non-U.N.: 1	3.6
War on Terror	12	9.1

thesis were correct, we would expect to see an increasing number of U.S. interventions whose principal purpose was the overthrowing of authoritarian regimes and installation of democracies, or the reversal of egregious human rights abuse. What we actually find is that while there was indeed a small increase in the number of humanitarian interventions – six in total – this figure is dwarfed by the number (and scale) of interventions prompted by the War on Terror. What is more, of these six interventions, five came in the 1990s and four of those in the first half of that decade. Since 1995, the U.S. has launched only two humanitarian interventions – Kosovo in 1999 (with NATO) and Libya in 2011. Indeed, Libya stands as the solitary example of humanitarian intervention in the first two decades of this century and this was not a unilateral mission, but a multilateral operation authorized by the U.N. Security Council (Resolution 1973 (2011)). Overall, the U.S. record on humanitarian intervention over the past two decades is identical to Sweden's. Meanwhile, the U.S. has been decidedly less active in humanitarian interventions than Nigeria. Neither Sweden nor Nigeria is invested in a strategy of liberal hegemony though, suggesting that this pattern of American intervention owes less to this grand strategy too. One final point also warrants mentioning – none of America's primarily humanitarian wars have proven "endless": most endured for less than a year and all were terminated long before the average duration of a civil war. Evidently, it is War on Terror wars that are endless, not humanitarian wars.

Finally, the realist thesis has difficulty explaining all the opportunities missed. If the U.S. has a grand strategy of using all available means to promote democracy and human rights, multilateral institutions (for Mearsheimer but not Walt), and market economies and free trade, why has it chosen, more often than not, to not do these things? Far from dependably promoting its liberal ideology, the U.S. has consistently supported illiberal regimes in the Middle East (notably Saudi Arabia), Central Asia (Azerbaijan, Uzbekistan, and Tajikistan), sub-Saharan Africa (Ethiopia and

Uganda), and elsewhere. And while the U.S. has occasionally launched humanitarian interventions, such as in Somalia and the Balkans, there is a much longer list of examples of inaction in the face of mass suffering, notably the Rwandan genocide in 1994, the Darfur genocide a decade later, and the killing of civilians by Saudi forces in Yemen more recently. And when Russia flattened Chechnya's capital, Grozny, posing the dilemma of whether to prioritize liberal values or hard-nosed realism, the U.S. chose the latter unequivocally. Most recently, in Syria, both the Obama and Trump administrations decided *not* to engineer regime change, preferring a strategy based on engineering a political agreement while satisfying U.S. interests in defeating ISIS and containing Iran. A determined liberal hegemon would surely have unseated Assad long before Russia became involved.

In its full context, the realist thesis thus rests on a relatively small number of cases, themselves driven by an exogenous shock – the 9/11 terrorist attacks. As late as August 2001, President George W. Bush's foreign policy was a relatively conservative one, driven by a view that American power should *not* be used to change other states and societies. That shifted on 9/11, first with the military campaign in Afghanistan and then the disastrous invasion of Iraq. Both interventions, so crucial to the narrative of endless wars, were caused more by 9/11 and the logic of the global War on Terror than by the logic of liberal hegemony. That is not to say that liberal hegemony was entirely absent, as Toby Dodge points out.[18] Confronted by the collapsed states its interventions produced, the U.S. administration embraced the principles of liberal peacebuilding, but that was hardly the principal consideration or public justification. Without those invasions and other elements of the War on Terror, the balance sheet would look very different indeed.

Yet, for all that, Table 6.2 shows that there was a dramatic increase in the average duration of U.S. wars initiated in the 2000s, and evidence from the study of civil wars shows that their average duration is increasing too. Endless wars may not be a uniquely American problem, but they are nevertheless a problem for international order and human well-being.

Explaining Endless Wars

The source of the world's contemporary endless wars lies in the changing nature of war and the attitudes that surround it, and principally the fact that war today is not primarily a contest between sovereign

[18] Toby Dodge, "Intervention and Dreams of Exogenous Statebuilding: The Application of Liberal Peacebuilding in Afghanistan and Iraq," *Review of International Studies*, 39, no. 5 (2013), 1189–212.

peers but a struggle between rulers and the ruled, with their bands of neighbors, allies, and sympathizers.

It has long been understood that civil conflicts tend to be longer and more vicious than international wars. They drive unvirtuous circles of despair. Often themselves responses to fractured societies, incomplete state-building, and authoritarian rule, civil war and violent rebellion tend to further erode the capacity of states and societies to manage their affairs peacefully, damage livelihoods, and fuel the passions of future conflict. Civil wars are often barbarous precisely because the combatants know each other so well.[19] What is more, because civil war is a violent contest over who should govern and on what terms, it entails a collapse of social order itself. For that reason, thinkers have long ascribed it as a special evil because it entails not just the violence of war but also a reversion to a pre-civilized state. Thus, Plato argued that there was no greater evil than civil wars, the Roman poet Lucan that civil wars brought no triumphs, and the Christian theologian Augustine that these were wars of a "worse kind."[20] The perils of civil war, and their attendant horrors, were well described by Thucydides in his account of the brutal fighting in Corfu: "Death raged in every shape; and, as usually happens at such times, there was no length to which violence did not go; some were killed by their fathers, and suppliants dragged from the altar or slain upon it; while some were even walled up in the temple of Dionysus, and died there."[21] The cause of this bloodletting was human nature and "the lust for power arising from greed and ambition." For Thucydides, the collapse of the polis precipitated a collapse of the body politic and with it the very possibility of order or justice. Thomas Hobbes also regarded civil war as a source of great calamity. Hobbes argued that the very purpose of political philosophy was to prevent "confusion and civil war," a task for which civil government "was ordained." From civil war, stemmed "all such calamities as may be avoided by human industry," including "slaughter, solitude, and the want of all things."[22] Its prevention, therefore, was among the first jobs of government.

But it is not just that civil wars tend to endure and be more brutal, it is also that efforts to redress them through international state-building and

[19] For example, Russell Jacoby, *Bloodlust: On the Roots of Violence from Cain and Abel to the Present* (New York: Free Press, 2011), p. 44.

[20] David Armitage, *Civil Wars: A History in Ideas* (New Haven, CT: Yale University Press, 2017), pp. 39, 46, 88.

[21] Thucydides, *History of the Peloponnesian War*, trans. Rex Warner (London: Penguin Classics, 1974), bk III, 82–85.

[22] Cited in Armitage, *Civil Wars*, p. 106.

modernization can provoke opposition and spur even more conflict. At the most macro level, Pankaj Mishra argues,[23] the rise of violent extremism that has prompted so much of the War on Terror is merely the latest phase of a backlash against modernization and globalization that goes back centuries and includes the anarchists who terrorized Europe in the nineteenth and early twentieth centuries. Mishra maintains that Western-style modernization uprooted traditional cultures and societies but failed to replace them with new locally grounded and legitimated ideas about how they should live. That was not much of a problem for those who benefited materially from the global transformation, but it created a reservoir of resentment among those who did not benefit, those on the receiving end of the ever-widening inequalities between rich and poor. Extremist ideologues – entrepreneurs of disenchantment – exploited this resentment to their own advantage.

The problem is compounded by the fact that civil war has a multiplier effect: civil war causes poverty, economic inequality, and sharp declines in wealth, which in turn make war more likely.[24] This is the "conflict trap."[25] Civil wars ruin economies, destroy social infrastructure such as housing, health, and education, and undermine faith in basic institutions, imposing intergenerational costs. The problems they introduce make civil war more likely in the future because wars occur more often when economies are weak, fragmented, and unequal, when access to rights and services are weak, and when societies are highly divided. Civil war thus puts societies into a dangerous downward spiral from which it is very hard (but not impossible) to escape. That is why when we look in detail at today's endless wars, in Afghanistan, Myanmar, Somalia, Iraq, Colombia, Nigeria, and the DRC, we see not a single contest stretching over decades but rather a pattern of punctuated violence involving different coalitions of actors interspersed with temporary truces and peace agreements. Studies show that the more parties there are to a war, the longer it is likely to last.[26]

[23] Pankaj Mishra, *Age of Anger: A History of the Present* (New York: Farrar, Straus and Giroux, 2017).

[24] For a summary of an extensive field of research see Jeffrey Dixon, "What Causes Civil Wars? Integrating Quantitative Research Findings," *International Studies Review*, 11, no. 4 (2009), 707–35.

[25] Paul Collier, *Breaking the Conflict Trap: Civil War and Development Policy* (Washington, DC: The World Bank, 2003); Håvard Hegre, Mokliev Nygård, and Ranveig Flatten Ræder, "Evaluating the Scope and Intensity of the Conflict Trap: A Dynamic Simulation Approach," *Journal of Peace Research*, 54, no. 2 (2017), 243–61.

[26] Dylan Balch-Lindsay and Andrew J. Enterline, "Killing Time: The World Politics of Civil War Duration: 1820–1992," *International Studies Quarterly*, 44, no. 4 (2000), 615–42; David E. Cunningham, "Veto Players and Civil War Duration," *American Journal of Political Science*, 50, no. 4 (2006), 875–92.

A further complicating factor is the increasing internationalization of civil war. The first two decades of the twenty-first century have seen a distinct trend toward greater internationalization as outside parties have become increasingly and more directly involved in civil war.[27] The evidence is fairly clear that external involvement in civil war tends to increase the duration of that war unless it is determinedly biased toward one of the parties and creates a preponderance of power that produces a decisive victory.[28] This is primarily because external interveners help sustain local parties in a variety of ways beyond their own exhaustion points, thus preventing one side from prevailing over the other with a decisive victory while also delaying the onset of a mutually harmful stalemate that might induce the parties to negotiate a settlement. Different forms of external intervention can have this effect. Most obviously, direct military assistance to a particular party can help it sustain the fight longer and inhibit it from negotiating a deal, while not necessarily giving it the wherewithal to achieve victory. For example, Russian and Iranian support for the Syrian government saved it from defeat, sustained it, and then helped it make territorial advances. Meanwhile, U.S., Turkish, Saudi, and Qatari support for Syria's rebels similarly helped sustain them. This is a global trend driven by the confluence of local wars and geopolitical competition. But impartial interventions, such as U.N. peacekeeping, can have this effect too, by managing local ceasefires and providing public goods that relieve the warring parties of the burden of fighting simultaneously across all fronts and of having to sustain their own civilian populations – and it is when rebels are not dependent on the civilians they control that atrocities become more likely.[29]

Nor is internationalization limited to direct involvement. The world is awash with arms and ammunition and simple, low-cost weapons easily traverse national boundaries. Another side effect is the contagion of war. Datasets show that, all other things being equal, a country whose neighbors experience war is far more likely to succumb to war itself than a country whose neighbors are at peace. The potential spread of civil war, one recent

[27] Sebastian Von Einseidel, "Civil War Trends and the Changing Nature of Armed Conflict," U.N. University Centre for Policy Research, Occasional Paper 10, 2017, https://collections.unu.edu/eserv/UNU:6156/Civil_war_trends_UPDATED.pdf.

[28] For example, Patrick Regan, "Third Party Interventions and the Duration of Interstate Conflicts," *Journal of Conflict Resolution*, 46, no. 1 (2002), 55–73.

[29] Jessica Stanton, *Violence and Restraint in Civil War: Civilian Targeting in the Shadow of International Law* (New York: Cambridge University Press, 2016).

study found, threatened neighboring states like nothing else.[30] There are a number of reasons why this might be. There is some evidence that violence itself is contagious – that violence begets violence, pushing societies and their neighbors into cycles of viciousness from which it is difficult to escape. Civil wars reach across borders as supply chains for arms, ammunition and soldiers, political kinship, and ideological affinities stretch across national boundaries. The civil war in Syria provides a good example. Not only did it give rise to renewed violence in Iraq, and potential violence in Lebanon, but it also drew in Turkey, Qatar, the United Arab Emirates, and Saudi Arabia on one side, and Iran, Iraqi Shi'ites, Hezbollah, and Russia on the other.

Thus, the changing nature of war goes a long way toward explaining the endless wars phenomenon. But there is another reason, attached to ethics and legitimacy, for the emergence of endless wars, relating to how our thinking about war influences our capacity to translate battlefield successes into sustainable political successes. Legitimacy matters because it influences how actors respond to external events. When actors believe a decision to be legitimate, they are more likely to voluntarily comply with it, even if it has negative impacts. More legitimate actions tend to receive more voluntary compliance and support; less legitimate actions tend to generate opposition. It is not difficult to see how this relates to endless war. If the decision on the battlefield becomes less legitimate, then it becomes more difficult for the victors in battle to translate military victory into political victory – something they must do before war can transition to peace.[31]

Over the past few decades, there has been a progressive tightening of what can be legitimately achieved by war. In particular, territorial conquest and colonization – once core rights of states – have been rendered illegal.[32] No longer can territories be added to a national homeland by force. Colonies cannot be seized and bent to the will of the colonial center. By 1945, almost the whole of the inhabited planet was divided into recognizable political entities, be they states, trusteeships, or colonies. For one to take over another is now forbidden. What this means in practice is that attempts to conquer another territory draw international censure, if not outright military

[30] Jacob Kathman, "Civil War Contagion and Neighboring Intervention," *International Studies Quarterly*, 54, no. 4 (2010), 989–1012.

[31] Cian O'Driscoll, *Victory: The Triumph and Tragedy of Just War* (New York: Oxford University Press, 2019).

[32] Tanisha Fazal, *State Death: The Politics and Geography of Conquest, Occupation and Annexation* (Princeton, NJ: Princeton University Press, 2007).

opposition, and, even if successful, cannot be translated into a new political fact. While this creates a powerful disincentive to initiating wars of aggression and thus contributes to the overall downward trajectory in the incidence of war, it also makes it more difficult to bring wars to a legitimate close. Russia's recent occupation of Crimea demonstrates the point nicely. Though dominant on the battlefield, prevailing without loss of life, and supported by a large majority of the population of Crimea, Russia has failed to translate its victory into a recognized legal fact. Crimea remains "occupied territory," with Russia subjected to economic sanctions and Crimea's citizens unable to travel freely around the world. Turkey faced a similar problem following its invasion of Northern Cyprus in 1974.

What is more, the rise of nationalism and self-determination has also made it more difficult to translate victory on the battlefield into political victory. If national groups refuse to accept defeat, as they have been increasingly reluctant to do in the age of nationalism, then the defeat of an army may result not in peace but in the emergence of insurgency, or irregular armed opposition. Whatever the merits of their cause, foreign invaders break the link between territory, people, and government and create a powerful incentive for "the people" to take up arms against them.[33] Napoleon found this to his cost in Spain – that nationalism breeds a deep sense of shared identity that pushes communities to resist invaders with all their might. During the peninsular war, the Spanish devised the perfect means of resisting a more powerful enemy – the *guerilla.* Guerrilla war involved the use of hit and run tactics, the avoidance of battle, and the use of civilian populations as cover. Though incapable of defeating the enemy outright, guerrilla tactics significantly increase the costs of occupation and reduce the payoffs, often wearing the more powerful side down. Because guerrillas avoid battle, the stronger side is unable to deal their enemy a decisive blow. Unable to distinguish the guerrillas from the civilian population, they sometimes turn to retribution against civilians. This, of course, is almost always counterproductive, since these patently unjust tactics only increases local support for the guerrillas. Meanwhile, Western governments confronted by insurgencies in Afghanistan and Iraq failed to note one of Clausewitz's principal lessons: that as policy becomes more ambitious, so war becomes more intense.[34]

33 Karma Nabulsi, *Traditions of War: Occupation, Resistance, and the Law* (Oxford: Oxford University Press, 1999).

34 Hew Strachan, *The Direction of War: Contemporary Strategy in Historical Perspective* (New York: Cambridge University Press, 2013), p. 16.

The use of guerrilla tactics by one or more of the parties is a feature common to all today's endless wars. The Taliban's insurgency in Afghanistan prevented the U.S. from translating its military success there in 2002 into a political success.[35] Likewise, it was post-2003 insurgencies in Iraq that helped produce an outcome antithetical to the war's original ambitions. This isn't new. Insurgency tactics were employed to good effect in numerous anti-colonial struggles and notably by the Viet Cong against the U.S. and by the Afghan *mujahedeen* against the Soviet Union. Armed with the tactics of guerrilla warfare, the conquered can make occupation and conquest prohibitively expensive. This is a lesson that almost every great power has had to learn at some point over the past two centuries.[36]

Armed secession is now also judged unlawful, creating a similar problem. During the nineteenth and early twentieth centuries, secession by force was one of the principal ways in which new states were born. Greece, Serbia, and Romania, for example, all owe their independence to wars waged against the Ottoman Empire (the latter by aligning with Russia in the Russo-Turkish war of 1877–78). Before that, the U.S. won its independence from Britain thanks to a secessionist war. All of these states were recognized as sovereign shortly after their military successes. There is nothing to suggest, for example, that the international community would not have recognized the Confederacy had it emerged intact from the U.S. civil war. In the early twentieth century, this right began to disappear as states recognized its potential for disorder. After 1945, the right disappeared entirely. Recognition and mutual agreement, not armed force, would be the engine of secession. Only if all the relevant parties agreed to secession would it be permitted. But in practice, parties rarely do agree – the partition of Czechoslovakia into the Czech Republic and Slovakia remains a rare exception. According to one study, the success rate of armed secessionist movements – never good – declined from 12 percent before 1945 to 2 percent afterward, the great majority of that 2 percent was tangled up in the process of decolonization.[37]

If we look at some recent cases, we can see just how difficult it has become to translate secessionist war into actual secession. The world's newest state, South Sudan, waged a decades-long war for independence

[35] Theo Farrell, *Unwinnable: Britain's War in Afghanistan: 2001–2014* (London: Vintage, 2017).

[36] Ivan Arreguín-Toft, *How the Weak Win Wars: A Theory of Asymmetric Conflict* (Cambridge: Cambridge University Press, 2005).

[37] Tanisha Fazal and Ryan Griffiths, "Membership Has Its Privileges: The Changing Benefits of Statehood," *International Studies Review*, 16, no. 1 (2014), 79–106.

from Khartoum. For most of that time, South Sudanese, not government forces, controlled most of what is now South Sudan. Yet secession only happened once the Sudanese government agreed to it in a comprehensive peace settlement. The same was true for Eritrea. It took thirty years of war (1961–91) to persuade Ethiopia to accept Eritrean secession, and only then was Eritrea granted international recognition of its sovereignty. Meanwhile, there are many examples of secessionist movements that did not succeed because their opponents chose not to relent. Most obviously, in the Middle East, the Kurds of Iraq, Syria, Iran, and Turkey are no closer to achieving statehood, despite decades of armed conflict and, in some parts of the region, significant periods of autonomous rule. The Igbo of Nigeria, Matebele of Zimbabwe, Chechens of Russia, Karen and Shan of Myanmar, Ossetians of Georgia, and Serbs of Bosnia are among the dozens of groups that have launched unsuccessful armed insurrections to secede from their states. Somaliland has done a remarkable job governing itself, yet its statehood will likely only be recognized if it succeeds in disassociating itself from Somalia by war. By making it more difficult for secessionist movements to succeed, and thereby putting downward pressure on the *incidence* of war, legal rules inhibiting secession may have had the perverse effect of encouraging war's increased *duration.*

Ending Endless Wars

How, then, are we to begin thinking of ways out of endless wars? Given that the problem of endless war is not an exclusively American problem, nor a by-product of its liberal hegemonic grand strategy (to the extent that it had one), there is little reason to think that dramatic adjustments to American foreign policy alone will make a significant impact. For the U.S., the question is one of whether, and how, to continue to combat violent Islamist extremists, and the record is a mixed one. For just as interventions in Iraq and Afghanistan failed to achieve their goals, the anti-ISIS coalition largely succeeded in its goals and the global campaign against al-Qaeda too largely succeeded in disrupting that organization's capacity to mount concerted and coordinated terrorist campaigns. The key issue in this context is to avoid collapsing quite distinct sets of issues into a single problem. For the U.S., the problem is one of translating battlefield success in the War on Terror into long-term sustained peace. This may involve finding a way of pursuing a War on Terror either without war at all or with more decisive war, perhaps through partnerships with reliable local actors, such as with the alliance with the Kurds that delivered victory over ISIS. Certainly, thinking through how to achieve peace in the War on Terror – a topic covered elsewhere in this volume – is central to remedying the American part of the endless wars problem.

More generally, we might identify two strategic lessons and some political food for thought arising from the preceding discussion. The first strategic lesson would be to avoid getting into endless wars in the first place. This is a question not just of whether or not to intervene, but of *how* to intervene too. Biased interventions directed toward helping a specific party win do not tend to result in endless wars, especially if the local partner enjoys legitimacy. By contrast, endless wars arise when objectives are vaguely couched in terms of achieving values (i.e., democracy) or amorphous goals (i.e., defeating terrorism rather than defeating a specific terrorist group). Where armed intervention is contemplated, its goal should be to bring about a decisive victory for a specific party as a means to achieving whatever secondary goals are intended. Where there is no clarity on these issues, little interest in achieving a decisive victory, or no stomach for allying with one or more of the local parties, intervention will likely precipitate a lengthy war. Thinking carefully about the peace to come must therefore be a core part of decision-making about whether and how to fight in the first place. When there are no good answers to those questions, intervention may be likely to yield endless war.

The second strategic lesson flows directly from the first. When contemplating intervention, ends should be carefully calibrated to means. Ambitious goals require ambitious means; if the available means are limited, then so too should be the goals. International intervention has the effect of prolonging civil wars because interveners typically provide enough support to keep the war going but not enough to bring it to a conclusion. That is because while they have an interest in the outcome, or in supporting or coercing a particular party, that interest is not sufficient to prompt a deep military commitment. Whether it is Iraq or Afghanistan, or Syria and Yemen, we see time and again external powers escalating and sustaining war but stopping short of committing the resources needed to achieve their stated goals. Across these and other cases, external powers have pursued ambitious goals with sometimes less than ambitious means with the obvious result being that those goals are not achieved. Since the time of Aquinas, just war theorists have included a prudential test in their evaluation of war's morality: will the proposed strategy and the means committed to achieving it deliver the desired effect? Where there is uncertainty about the answer to that question, there are good reasons to doubt the prudence of military action and every reason to expect that it might result in an endless war.

At the heart of the problem of the endless war, though, lies the fact that different societies have different interests and values, and that war remains the *ultima ratio* of disputes between them, the ultimate determinant of which side gets to have its way. Indeed, as Hannah Arendt explained, "the chief reason warfare is still with us is neither a secret death wish of the human species, nor an irrepressible instinct of aggression, nor, finally and more plausibly, the serious economic and social dangers inherent in disarmament, but the simple fact that no substitute for this final arbiter in international affairs has yet appeared on the political scene."[38] "For all its evil," George Orwell explained, war is "an unanswerable test of strength."[39] Yet because it is an imperfect arbiter, there must always be an accommodation between victor and vanquished to sort out what is just, who governs what and where, and on what terms. Order requires that the outcomes of these tests of physical strength be recognized by other members of society. That is why, as O'Driscoll reminds us, early jurists like Gentili and Grotius admitted that victory in war, even unjust wars, established legal rights, rights that others must recognize in the interests of order.[40] Ultimately, in other words, force must always reach an accommodation with law and justice.

Even if major powers become smarter in deciding when and how to intervene abroad, endless war will remain with us for as long as war serves this function. But war is an unjust and unreasoning arbiter, a contest of strength and guile, not of rights. War tells us who is stronger, not what law, justice, and reason have to say about the case. War, in other words, turns contests over justice into contests of strength. Addressing this problem involves not only thinking about the politics of intervention but also about the deeper structural reforms needed to make war a less compelling political arbiter and to replace it with other authoritative ways of managing existential crises and disputes.[41] This ought to involve not just careful thought about the political and moral procedures for negotiating peace out of war, but also heightened attention to figuring out ways of arbitrating disputes and satisfying rights without the need for war.

[38] Hannah Arendt, *On Violence* (London: Harcourt, 1970), p. 5.

[39] George Orwell, *The Lion and the Unicorn: Socialism and the English Genius* (London: Penguin, 1941 [1970]), p. 37.

[40] O'Driscoll, *Victory*, pp. 90–107.

[41] For example, Alex J. Bellamy, *World Peace (And How We Can Achieve It)* (New York: Oxford University Press, 2020).

Appendix: *Ongoing Wars 2020 (ongoing conflicts with at least 500 deaths in 2020)*

	Start year	Location	Belligerents	Casualties total	Casualties 2020	Duration (years up to Dec 2020)
1	2020	Ethiopia	Govt. Ethiopia vs. Tigray	1,261	1,261	0.5
2	2017	Cameroon	Govt. Cameroon vs. Anglophone	3,000–4,000	524	2.5
3	2017	Mozambique	Govt. Mozambique vs. Islamist insurgents	3,000–4,000	1,662	2.5
4	2015	Syria/Iraq	ISIS vs. Govt. Iraq, U.S. coalition, Kurds, Syrian opposition, Russia	90,000–200,000	4,000–8,000	4.5
5	2014	Yemen	Houthis vs. Govt. Yemen/Saudi/allies vs. AQ/ISIS	100,000–150,000	18,141	6
6	2014	Libya	Transitional govt. + allies vs. Haftar + allies	20,000–45,000	1,480	6
7	2012	Mali	Govt. Mali + allies (U.N., U.S.) vs. Tuareg + AQIM + ISIS	11,000–13,000	2,400	8
8	2011	Syria	Govt. Syria (+ Russia, Iran, and allies) vs. opposition, Kurds + allies	350,000–600,000	7,625	8.5
9	2011	South Sudan	Govt. South Sudan (SPLA) vs. SPLM-IO	380,000–400,000	2,240	8.5
10	2011	Egypt (Sinai)	Govt. Egypt vs. Islamists (inc. ISIS)	5,000–6,000	522	9
11	2009	Nigeria	Boko Haram vs. Govt. Nigeria	50,000–60,000	5,110	11
12	2006	Mexico	Govt. Mexico (+ allies) vs. drug cartels	41,000–150,000	7,540	14
13	2003	Iraq	U.S. + allies (inc. Govt. Iraq) vs. insurgents	300,000–1,200,000	2,380	16.5

14	2003	Darfur	Govt. Sudan vs. JEM, SLA, SRF + others	200,000–300,000	868	16.5
15	2002	Maghreb	AQIM vs. Govts. of Algeria, Tunisia, Mali, Burkina Faso, Chad, Mauritania, etc. + allies (inc. U.N., U.S.)	10,000–20,000	5,845	17.5
16	2001	Afghanistan	Taliban/AQ vs. U.S. + allies	280,000–500,000	17,631	18.5
17	1998	Nigeria	Communal conflicts (various)	15,000–20,000	600–800	22
18	1996	DRC (Ituri, Kivu, ADF)	Lendu + allies vs. Hema + allies (inc. Govt. DRC), M26/7 vs. Govt. DRC + allies vs. FDLR vs. NDC-R, Govt. DRC vs. ADF	60,000–100,000	5,460–6,460	24
19	1991	Somalia	Transitional govt. + allies (inc. Kenya, U.S.) vs. al-Shabaab, ISIS + allies	400,000–600,000	2,759	29
20	1988	Azerbaijan	Azerbaijan vs. Armenia	20,000–30,000	4,300	31.5
21	1984	Turkey	Govt. Turkey vs. PKK	30,000–70,000	519	35.5
22	1964	Colombia	Govt. Colombia vs. FARC	200,000–250,000	693	56
23	1948	Myanmar	Govt. Myanmar vs. Karen, Shan, Kachin, Rohingya, Rakhine	130,000–210,000	570	72

CHAPTER 7

Forever Wars
Time and Value in War

David Rodin

In this chapter, I want to examine some of the underappreciated implications of time in the ethics of war by considering a particular limiting case of conflicts – wars that extend so far in time that we may consider them "endless wars." What are endless wars? Clearly, no wars are strictly without end. One day the earth will perish and everything with it, so all things come to an end. Nonetheless, there is a more limited conception of endless wars that will be a useful tool to explore a set of difficult issues raised by the role of time in the ethics of war. Endless wars, as I shall use the term, refer to conflicts that continue over a highly extended period because they have no foreseeable and determinate path to victory.

Endless wars can be divided into two different classes. The first is conceptually endless wars. These are conflicts in which the war goals are defined in such a way that no combatant could ever achieve them. The Global War on Terror is arguably conceptually endless in this sense. No foreseeable military strategy could ever have the effect of bringing about an end to "terror," therefore any combatant committed to the War on Terror is committed to prosecuting a war with no ending point. A second class is practically endless wars. These are conflicts in which victory cannot be achieved within the normal planning cycles of political and bureaucratic institutions. The wars in Afghanistan, Iraq, and, before them, the war in Vietnam are practically endless wars. Although victory in a classical sense is not impossible, there is no foreseeable and determinate path to victory within normal planning cycles. Lower-level conflicts may also exhibit the features of a practically endless war. The Israeli-Palestinian conflict, considered an ongoing state of hostilities, has continued for more than 50 years. Moreover, the use of special forces or other clandestine use of force can also be endless in this sense.

I propose to consider three problems raised by endless wars. First, do endless wars necessarily fail the proportionality and reasonable prospect of success conditions within *jus ad bellum*? Second, how do we deal with the problems of indeterminacy in assessing costs and benefits over long time frames? Third, should future costs and benefits be discounted compared to present costs and benefits in proportionality judgments? If so, what effect does this have on the ethics of war?

Endless Wars and *Jus ad Bellum*

Why might we think that endless wars necessarily fail the proportionality and reasonable prospect of success conditions of *jus ad bellum*? The argument is easy enough to construct: Just wars have objectives that are specified by their just cause. The just cause is achieved by attaining victory in the war. A war is proportionate if the costs required to achieve victory in pursuit of a just cause are proportionate to the moral value of the just cause. Similarly, war has a reasonable prospect of success if there is a clear and viable path to victory. But endless wars are precisely those wars for which no foreseeable and determinate path to victory exists. Therefore, endless wars are necessarily disproportionate and without reasonable prospect of success, and for these reasons are, all things considered, unjust from an *ad bellum* perspective.

This argument is attractive, but it is also wrong. The mistake it makes is to equate the achievement of a war's just cause with the achievement of victory. For certain forms of just wars the achievement of the just cause and the achievement of victory are identical, but this is not invariably the case. It can be possible to achieve a just cause without achieving victory if the just cause consists of frustrating an enemy's unjust intentions. This may be the case in wars of self-defense or humanitarian intervention, where the just cause consists of preventing the enemy from achieving their purpose of committing an atrocity or extinguishing the sovereignty of a political community. In such cases, it may be possible to achieve the just cause simply by denying victory to the enemy, even if there is no clear path to military victory against them. Such conflicts would inevitably inflict a sustained series of moral costs, but they may still conceivably fulfill the *ad bellum* and *in bello* proportionality requirements as demonstrated by the time series chart below.

In Figure 7.1 the black bar represents the moral costs of an ongoing conflict and the gray bar represents the benefits (for example frustrating an atrocity or maintaining sovereignty) that accrue in a particular time period. This distribution of costs and benefits could be extended indefinitely in time and – so long as benefits continue to exceed costs period-by-period – the war as a whole would be *ad bellum* proportionate. Assuming that the just cause consists of frustrating the enemy's unjust threat of harm, such a conflict would also meet the reasonable prospect of success condition of the *jus ad bellum*.

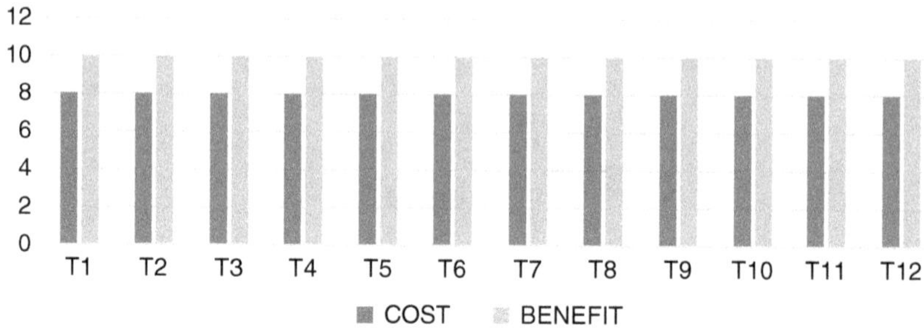

Figure 7.1 Constant cost/benefit time sequence

Proportionality and Indeterminacy

To be clear, endless wars, because they have no determinate path to military victory and therefore extend the costs of conflict into the foreseeable future, are highly likely to fail the *ad bellum* proportionality and reasonable prospect conditions – but do they not necessarily do so. However, there is an additional problem with endless wars. How can we determine whether a war that extends well beyond normal institutional and bureaucratic planning cycles is proportionate or not? Consider a conflict with the following profile of costs and benefits:

The total costs and benefits in Figure 7.2 over the period T1–T12 are identical to those in Figure 7.1. However, instead of exhibiting a constant relationship between costs and benefits for each time period, Figure 7.2 shows a variable cost-benefit profile. This represents a more realistic picture of the cost profile of military conflict. In line with many other forms of long-term projects, benefits tend to lag costs. So, a period of heavy activity (generating high costs) yields little immediate tangible benefits (period 1). This is followed by a period in which benefits start to materialize and costs can be reduced (period 2). Sometimes committed resources are reduced so much that benefits start to drop off and a new cycle of cost increase followed by lagging benefits occurs (periods 3 and 4).

The problem is not that a variable profile of costs and benefits cannot be in totality value positive – the data presented in Figure 7.2, as in Figure 7.1, shows an aggregated surplus of benefits over costs of 25 percent. The problem is how can we know that this is the case, when the dataset is extended indefinitely into the future? Whether our assessment shows the conflict to be value positive or value negative clearly depends on which period we examine. Periods 1 and 3 show value negative (disproportionate) outcomes

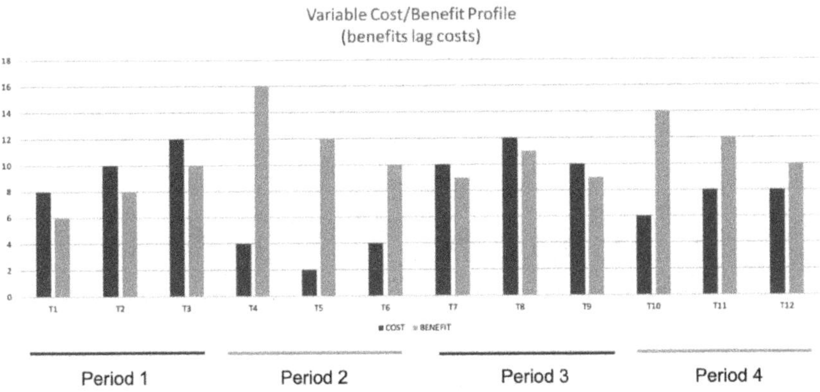

Figure 7.2 Variable cost/benefit time sequence

and periods 2 and 4 exhibit value positive (proportionate) outcomes. This would seem to create a significant difficulty for determining the *ad bellum* proportionality of endless wars. Wherever we are in the course of active conflict, we cannot be sure that we are not simply seeing a local anomaly within a larger trajectory.

This is the problem of indeterminacy. It affects our ability to assess the proportionality of a conflict that is already in progress. But the problem is considerably worse for those who must make the decision of whether to enter a state of war before the conflict has begun. *Ad bellum* considerations require a decision-maker to assess the likely costs and benefits of the action and the likely prospect of successfully achieving the just cause. But for a conflict that has no viable and determinate path to victory, they must attempt to estimate likely costs and benefits over an indeterminate and extended future period. Not only are these costs and benefits likely to be variable over time (as in the chart above), but forward-looking assessments can only ever be estimates and therefore subject to a risk of error. This has always been a problem for forward-looking *ad bellum* judgments since the task of estimating the likely effects of a large-scale, highly complex, and adversarial process like war is fiendishly difficult. Moreover, there is a powerful psychological tendency for planners and decision-makers to unconsciously wish for the best, underestimating risks and overestimating likely benefits. This is the reason that many wars last much longer and yield far more meager benefits than expected. The problem with endless wars is that they have the potential to hugely amplify the indeterminacy of forward-looking proportionality judgments arising from erroneous assessment. This is because predictive assessments in the far future are inherently less

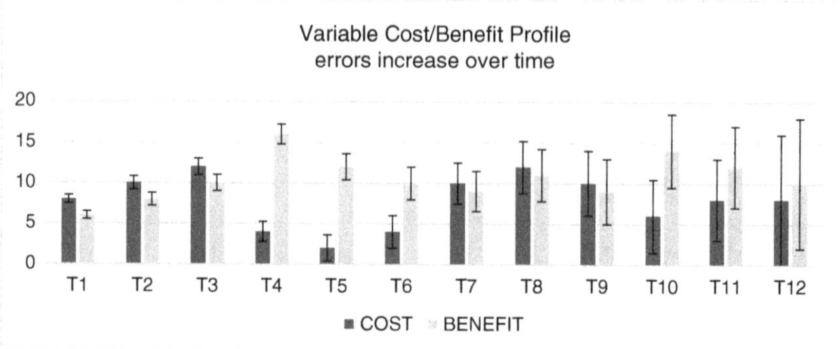

Figure 7.3 Variable cost/benefit time sequence with margins of error

reliable because of the potential for unforeseen and random occurrences to change the course of events and thus the ethical distribution of costs and benefits. Thus, the forward-looking distribution of costs and benefits will look more like in Figure 7.3, with error rates increasing over time.

There is a way to manage indeterminacy in decision-making by allocating a presumptive burden of proof. The effect of indeterminacy on forward-looking *ad bellum* decision-making will therefore depend on where the burden of proof lies in proportionality judgments. If one believes that otherwise justified defensive action is morally permissible unless its consequences can unambiguously be shown to be disproportionate, then the indeterminacy of outcomes in endless wars will have a strongly permissive effect – liberating decision-makers to choose conflict even when outcomes are ambiguous. If, on the other hand, one believes that otherwise justified defensive actions are morally impermissible unless their consequences can be unambiguously shown to be proportionate, then the indeterminacy of outcomes will have a strongly restrictive effect – creating a strong presumption against entering conflicts that have no clear and determinate path to victory.

Which is the correct presumptive burden of proof to impose in circumstances of indeterminate outcomes? A strong case can be made that the correct interpretation is the second, restrictive, interpretation of the burden of proof. Two considerations will be relevant. The first is that when we choose to enter a state of war, we impose costs as a result of our own action or agency, whereas the benefits of war consist of thwarting the harmful acts of others. As the doctrine of doing and allowing reminds us, we have a stronger moral reason to avoid inflicting harms on others through our own agency than we have to prevent harms arising from the agency of others.

This consideration will push us toward a more restrictive interpretation of the burden of proof in proportionality judgments.

Balanced against this is the fact that we have weaker moral reasons to avoid harming those who are potentially liable to the harm, than those who are not. When we choose to go to war, at least some of the harm that our action inflicts falls on those who are potentially liable to this harm (the enemy unjust combatants). In addition, the unjust harm that we avert through prosecuting a just war would otherwise have fallen entirely on persons who are not liable to suffer this harm. This consideration will favor the permissive interpretation of the burden of justification – however, it will do so less strongly because its valence will be mixed. While some of those we harm through war are potentially liable to this harm, the war will also inflict harm on those who are not liable to be harmed (the just combatants of our own forces, and the just and unjust combatants on both sides of the conflict). Taking both considerations into account it would seem reasonable that the burden of proof for proportionality should be structured to not permit military action unless the consequences can unambiguously be shown to be proportionate.

The third issue raised by endless wars that I wish to discuss is the impact of time on forward-looking value assessments. Proportionality requires comparing the moral value of costs and benefits, but when these costs and benefits are distributed over long periods of time – as they are in endless wars – we cannot assume that they will have a constant moral weight for the purposes of proportionality. The variance of value assessment over time is a highly controversial issue, addressed by a large body of literature in both philosophy and economics.[1] I cannot hope to fully resolve these issues here, but my objective is to show that time effects are relevant to wide proportionality judgments and indeed are a very serious consideration that will have potentially significant implications for the conduct of long-duration wars. To my knowledge, these issues have not been discussed in the ethics of war literature.

Time-Relative Value and Proportionality

The costs (and also the benefits) of war can be broadly divided into two kinds – colloquially characterized as "blood and treasure." Let us begin with the effect of time on treasure, which is far the easier of the two.

[1] For a useful summary see David Weisbach and Cass Sunstein, "Climate Change and Discounting the Future: A Guide for the Perplexed," *Yale Law & Policy Review*, 27, no. 2 (2009), 433–57.

Financiers and economists have long understood that the present value of money received at different points in the future is not constant, but rather diminishes the further in the future that money accrues. For example, $100 received in one year's time has less value than $100 today. The reason for this is opportunity costs. If I have $100 today, I can invest that money at a rate of return and in one year I will have more than $100. Therefore, money received today is worth more. To put this more precisely, the present value of future monetary gains or losses must be discounted. The effect of time on the present value of money has been formalized to allow us to calculate the current value of future monetary gains or losses. The formula is:

$$PV = FV / (1 + r)^n$$

where PV is the present value of a future financial gain or loss; FV is the future value of the gain or loss; r is the available rate of return; and n is the number of time periods.

This is one of the fundamental formulas of finance. Without it, it would be impossible to rationally assess the financial attractiveness and viability of long-term projects. Whenever investors allocate capital, or managers or policymakers invest in a capital expenditure project, they use the present value formula to determine whether the value of future returns exceeds the upfront investment.

Figure 7.4 shows the present value of 100 currency units over time on the assumption of two different discount rates – 5 percent, which is the approximate long-term rate of return of U.S. equities, and 2 percent, which is the approximate long-term rate of return of U.S. treasuries.

There are several things we can note from this chart. First, the effects of time discounting on the present value of money are large. The present value of a monetary cost or benefit that accrues after 10 years is just 82 percent of a nominally identical cost or benefit that accrues immediately – assuming a 2 percent discount rate. On a 5 percent discount rate, the present value falls to 61 percent.

This has a number of significant implications for how we should think about the monetary costs and benefits of endless wars. First, it creates incentives to structure military operations in such a way as to bring forward, or front-load financial benefits, while pushing back or deferring financial costs. One obvious way to do this is through the use of debt to defer the costs of military expenditure – and war and debt financing have a long historical interdependence. But it may also create moral reasons to take differing approaches to the conduct of military operations. For example, combatants

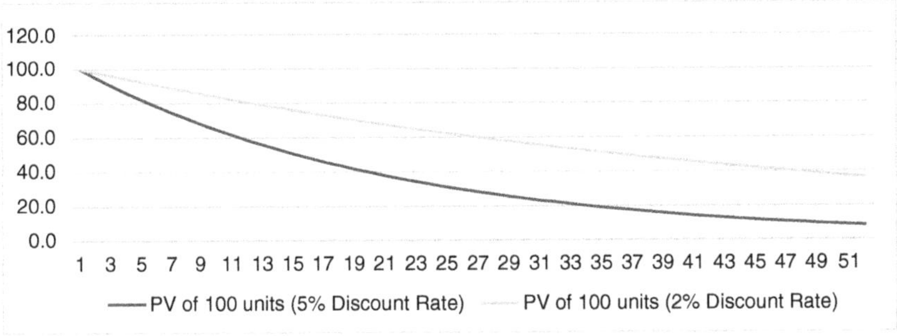

Figure 7.4 Present value of 100 value units over time

pursuing a just cause may choose to prioritize operations to protect productive assets like arable land and industrial capacity early in the conflict so as to maximize their present value in *ad bellum* proportionality calculations. Equally, they may choose to delay destructive kinetic operations against enemy plants and equipment until later phases of a conflict, so as to minimize the present value of these costs for the purposes of proportionality. These implications follow straightforwardly from the observation that financial costs and benefits are morally significant impacts of war that must be counted in *ad bellum* and *in bello* proportionality calculations. Still, it is somewhat disconcerting to think that the balance between a war being just and unjust may conceivably hinge on actions that are essentially matters of financial planning and engineering – the availability of debt financing at a lower rate than the effective discount rate, or the ability to structure operations so as to push back financial costs and front-load benefits.

Second, the discounting of future value will help to mitigate the indeterminacy problem discussed above. We have discussed how forward-looking *ad bellum* decisions require estimating costs and benefits far into the future. These costs and benefits are not only distributed variably, but our estimation errors will increase the further out in time we look. Temporal discounting mitigates this problem because, although our ability to accurately assess costs and benefits diminishes with time, so too does the contribution made by those costs and benefits to *ad bellum* justification. After a certain period, additional costs and benefits make only a negligible contribution to proportionality judgments on a present value basis.

Third, note the importance of our assumptions about the relevant discount rate to our understanding of the present value of future costs and benefits. The differences in present value between assuming a 2 percent and a 5 percent discount rate is very significant indeed. In many policy areas where present value discounting has been used, significant debate centers on the appropriate rate of discounting.

Discounting Human Costs

So much for war's costs and benefits in terms of treasure. What about the costs of war in terms of "blood," which we may take to include not only death and injury but also displacement, trauma, loss, and insecurity? Should nonfinancial costs also be discounted for time, so that a death or injury that occurs in the future is weighed less for forward-looking proportionality calculations than one that occurs today? If this were the case, it will have an even greater strategic and operational impact on the permissible waging of war. It will create a moral incentive to concentrate operations that protect populations early in a conflict, while deferring counterforce operations or those that impose significant risks of collateral damage until later in the conflict. It will also dramatically alter the moral status of certain kinds of weapons. It will increase the moral attractiveness of weapons – like landmines and cluster munitions – that produce immediate operational effects, but whose costs can accrue far into the future. It will dramatically decrease the moral attractiveness of tactics – like siege and boycott – whose costs are immediate but whose effect is only realized over long periods of time.

The notion that not only financial goods, but also human welfare should be discounted over time derives from the work of Frank Ramsey, who developed the notion of a "social discount rate" in his celebrated 1928 article "A Mathematical Theory of Saving."[2] Ramsey himself thought that the social discount rate should be set at zero – that is to say that the present value of future nonmonetary goods should not be discounted. But later generations of economists have taken it as axiomatic that future welfare goods should be subject to a positive social discount rate, normally taken to be comparable to the investment discount rates commonly used in the present value formula. This assumption has been hugely influential in the debate on climate change policy. Two of the most prominent climate commentators – Nicholas Stern and William Nordhaus – reach dramatically different conclusions on whether we should take strong action now to avert climate change, although both agree on almost all the basic science and outcome modeling. Stern believes we should take immediate

[2] F. P. Ramsey, "A Mathematical Theory of Saving," *Economic Journal*, 38, no. 4 (1928), 543–59.

and aggressive action to avert climate change because he assumes a social discount rate for future welfare of 1.4 percent, while Williams believes we should take far fewer actions today because he assumes a discount rate of 5.5 percent. The social discount supposition also appears to be validated by empirical studies of revealed preferences, which consistently show that people have a tendency to value future welfare less than present welfare.[3]

The supposition that nonfinancial goods and harms should be discounted over time has, however, been vehemently rejected by many philosophers. Sidgwick, Rawls, and Parfit all argue against "pure time discounting" – the discounting of a good simply on the basis of its position in time (as opposed to the probability of its occurring or the wealth of the recipient of the good, both of which can be correlated with the temporal positioning of a good). Their arguments are twofold: First, they argue that social discounting violates a principle of ethical neutrality and is hence irrational. As Rawls says: "rationality implies an impartial concern for all parts of our life. The mere difference of location in time, of something being earlier or later, is not in itself a rational ground for having more or less regard for it."[4] Second, they point out that assuming a social discount rate leads to fantastic and ethically abominable conclusions. For example, Parfit points out that, assuming a social discount rate of 5 percent, one statistical death next year counts for more than a billion deaths 400 years from now.[5]

What is more, even the staunchest advocates of social discounting only appear to be ready to apply the principle to the optimal distribution of goods. Examples include debates about the appropriate savings and investment rate for a society and climate change policy where the question is what burdens present generations should bear in order to avert harm to future generations. As far as I am aware, no one has attempted to apply social discounting to considerations of proportionality in defensive rights, where our question is the potential justification of harming one group of people in order to or in the course of benefiting another.

Nonetheless, I believe that it is not irrational to discount future nonmonetary costs and benefits – indeed far from violating the principle of ethical neutrality, I believe it is implied by it. Moreover, I believe that discounting can be relevant to proportionality in situations of interpersonal

[3] See, for example, Maureen L. Cropper, Sema K. Aydede, and Paul R. Portney, "Rates of Time Preference for Saving Lives," *American Economic Review*, 82, no. 2 (1992), 469–72.

[4] John Rawls, *A Theory of Justice* (New York: Oxford University Press, 1999), p. 259.

[5] Derek Parfit, *Reasons and Persons* (Oxford: Oxford University Press, 1984), p. 482.

defense and war. It will be helpful to approach the argument in stages, looking first at the tradeoff between monetary expenditure and future harm, and then extending the argument to wide proportionality, the form of proportionality used in moral lesser-evil reasoning.

Consider the question – very common in war – about what level of financial expense we are morally obligated to bear in order to avert a risk of harm to nonliable persons. A classic example would be an attack on a military target that can be undertaken using either conventional bombs or using smart bombs that carry a lower risk of collateral harm to noncombatants but are more expensive. One potential answer to this question is that if the smart bombs exist, then using the conventional bombs would inflict unnecessary harm on noncombatants and is therefore impermissible. In other words, one is always obligated by the principle of necessity to use the least harmful means available. But this answer cannot be right. If it were, then there would be no limit to the financial costs we would be required to bear in order to prevent side-effect harms. The relationship between financial cost and harm would not be a tradeoff, but an overriding objective. So long as there are marginal safety improvements to be realized, this imperative would swallow all other elements of defense spending – indeed of all public and private spending, crowding out any expenditure on recreation, culture, education, and other goods not directly contributing to safety improvements.[6] In order to avoid this conclusion, we must deploy some conception of a morally reasonable cost required to prevent a side-effect harm like death.

Policymakers throughout the world deploy a conception very like this in order to assess the merits of proposed expenditure on safety and environmental protection programs. This measure is usually called the value of a statistical life (VSL) and it is used to define a reasonable threshold of expenditure in order to save a single statistical life. In the United States, most agencies utilize a VSL of approximately ten million dollars. VSL can be conceptualized and calculated in different ways, most often either by studying the "revealed preferences" of persons (for example by asking them in surveys how much they would be willing to pay to avert a future harm) or by calculating the expected earnings potential of an average person through their life. Neither methodology provides a good measure for our concern, which is the financial costs we are *morally* required to bear in order to avert a harm such as a death. For this, we will require a different

[6] I will return below to the question of why this implausible conclusion is not entailed by the principle of necessity.

kind of measure. To distinguish it from the more usual definitions of VSL used by economists and policymakers, let us call this measure the morally reasonable value of a statistical life (MVSL). This can be defined as the currency value that it is morally reasonable to expend to save one statistical life. I will say more about how we can flesh this definition out below. For the moment, we need only note that in order to resolve questions such as the moral decision over whether to use conventional or smart munitions in an operation, we require some conception of a morally reasonable assumed financial value of a statistical life – MVSL.

We will first look at how such a measure can be utilized in decision-making involving contemporaneous costs and benefits and then consider the effects of time variance. Suppose that an otherwise permissible military operation is expected to cause 100 unintended deaths to nonliable persons, and that all of these deaths could be avoided by switching to smart munitions at a cost of $10,000. If the assumed MVSL is $100, then it would be morally required to switch 100 percent of the munitions to smart bombs (the combined value of MVSLs saved is equal to the additional expenditure required to save those lives). However, if our assumed MVSL were 80 currency units, then it would not be required to use 100 percent smart bombs – it would be permissible to use a mix of 80 percent smart munitions and 20 percent conventional, because that would be consistent with the morally reasonable requirements for financial expenditure to avert harm.

Let us turn now to time-variant expenditure decisions. Imagine a variation on the above case in which an otherwise permissible military operation requiring uranium tipped munitions will cause 100 unintended deaths to nonliable persons through radiation poisoning in ten years' time. These deaths can be averted using alternative munitions that are specially shielded at an additional cost of $10,000. Assume that, as above, the MVSL is $100. Assume, also, that planners are committed to a principle of intertemporal equality so that the MVSL of all lives should be identical, no matter where in time they are situated (for simplicity we assume a zero-inflation environment). What proportion of shielded to unshielded munitions should the planners deploy?

One might think that the answer is identical to our first case above: the planners must use 100 percent upgraded shielded munitions ($100MVSL x 100 lives saved = $10,000 [the cost of upgrading all the bombs]). However, this is incorrect. Spending $100 today to save a life ten years from now implies that the MVSL of those future lives is higher than $100 – which violates the principle of intertemporal equality. We can derive this from a simple correlate of the present value formula that enables us to calculate a future value on the basis of present value: $FV = PV(1 + r)^n$. Using this formula, we can infer that the implied MVSL of the lives saved in the future by the policy of spending $100 today is actually $122 (assuming a 2% rate of return). The correct answer is that if we wish to respect intertemporal neutrality by attributing the same moral value to all lives no matter where in time they are located, then we must deploy a discount rate. If we assume the discount rate to be 2 percent, we arrive at a mix of 82 percent shielded munitions and 18 percent unshielded ($100 MVSL per life saved in 10 years discounted by 2% yields a present MVSL of $82). The lesson from this example is that refusing to deploy a social discount rate does not treat all lives as having equal value, rather it treats future lives as having greater value than present lives. The further into the future we look, the greater this inequality of attributed value becomes.

Now this argument might feel like a sleight of hand. Whether it is appropriate to apply a discount rate to nonfinancial goods and harms is what is under dispute, and yet this argument has assumed the validity of applying a discount rate to reach its conclusion. That sounds like it may be viciously circular. But it is not. Recall that the MVSL is a financial unit of value. It is defined as the dollar amount that it is morally reasonable to spend in order to save a statistical life, and as a financial unit of value, it clearly has time variability. If we accept the validity of a measure such as MVSL, then we are committed to discounting. The present value formula is not an association or correlation relationship between different currency values – it is an equation. If we are committed to applying the same MVSL value to all lives no matter where in time they lie, then we must discount present spending required to save future lives.

But perhaps the problem lies with the notion of an MVSL itself. Is it coherent to attribute a currency value to a moral good like a human life? Aren't those radically incommensurable forms of value? Isn't it the attempt to put a monetary price on a moral value that is leading us astray here? I don't believe so. When we deploy the notion of an MVSL in moral reasoning, we should not see this as reducing the moral value of a human life to a monetary value – which would indeed be objectionable. Rather we are doing the opposite – we are unpacking the implicit moral value of money. We do this in two ways. First, money is a store of moral potential. This is because money can be used to realize moral goals and moral values. When we spend money for shelter and sustenance, for leisure and travel, to buy a gift for a loved one, to create a cultural artifact, or to hedge against a future risk, we transform the moral potential of money into real moral value.

The second way we can unpack the moral value of money is by reflecting on how possessing a monetary asset provides us with optionality – partly because it has the potential to be transformed into such a multifarious array of moral goods. Money, therefore, contributes to the moral value of enhancing our domain of free agency. What we are doing when we employ MVSL-style reasoning is to crystallize in a monetary measure the comparison between the protection of a life and the moral value of these underlying agency and potential goods. In some ways, this can be seen as a recasting of the traditional methodology for calculating VSL as an aggregate of potential future earnings. In this approach, we attempt to quantify the *monetary* potential of an average human life (in terms of future earnings). The MVSL, in contrast, attempts to quantify the *moral* potential of money.[7]

Understanding MVSL in this way enables us to see that problems concerning how much we are obligated to spend to save a life (like the smart bomb/conventional bomb problem) are actually just a form of a wide proportionality problem. The moral comparison we are making with the help of an MVSL measure is not between the moral value of a human life and the financial value of a sum of money. Rather it is a wide proportionality comparison between the value of the life and the value of the underlying moral potential implicit in the money. To be sure this requires making difficult value comparisons between widely differing forms of goods (life versus a wide array of moral goods that money can be exchanged for). But these comparisons are no more problematic than the ones that are required of standard wide proportionality judgments in war and self-defense (for example comparing the value of national sovereignty with the disvalue of death and bodily injury).

Understanding MVSL as a subspecies of wide proportionality also enables us to understand why the principle of necessity does not require us to spend limitless funds to prevent avoidable deaths. We noted above that an interpretation of necessity that required us to do so would be implausible as it would effectively require us to divert all public and private funding to safety programs. The current discussion of the moral underpinnings

[7] To say that money has an implicit moral value does not commit us to either a conceptual or an empirical claim that money always produces moral goods. Of course, money need not be used to realize moral goods and may often yield moral harms, and yet money should still be viewed as having an implicit moral value for the reasons given. An instructive analogy is with the capacities of autonomy and agency – to which monetary goods are closely connected. My possession of the capacities of autonomy and agency can have moral value, even though my exercise of those capacities can sometimes create harm both for myself and others.

of MVSL explains why this is not entailed by the principle of necessity. The necessity condition, properly specified, states that an action producing harmful side effects is necessary if there is no other *less morally costly* way of achieving the same objective. When the value of the MVSL is properly set, it formalizes this condition. The MVSL *just is* the point of expenditure, above which spending more to save a life would destroy greater moral value than the value of the life saved. Harms that could be avoided through expenditure above this level are therefore not unnecessary in the morally relevant sense.

I have argued that judgments over how much money we are required to spend to save a life are a form of wide proportionality judgment and that MVSL assessments must be discounted over time. Can we also conclude that values in wide proportionality judgments more generally should also be discounted over time? I believe that we can. Wide proportionality problems and morally reasonable expenditure problems can be transposed on each other. Consider the case of a military commander planning an operation and possessing a budget for future expenditure of $10,000. They can spend this sum on a consignment of precision-guided munitions that will prevent 110 unintended deaths among enemy civilians. Or they can spend it on a consignment of body armor for their own forces, which will prevent 100 deaths (I assume the consignments cannot be split, both groups are nonliable, and I ignore any agent-relative obligations arising from special relationships or duties of care). This is a wide proportionality problem that can be stated and resolved in MVSL terms. Assuming each life has an MVSL of $100, upgrading munitions yields combined MVSLs of $11,000, whereas purchasing body armor yields combined MVSLs of $10,000. Both represent a morally reasonable expenditure to save life, but the former yields a greater total MVSL. For this reason, it would be disproportionate and therefore impermissible to spend the budget on body armor rather than precision munitions.

Now consider a variant where the commander has the option to spend the budget on body armor that will save 100 lives now, or on shielded munitions that will save 110 lives in 10 years' time. Here the MVSL analysis is very different. Buying the body armor will yield a combined MVSLs of $10,000 but buying the shielded munitions will yield a present combined MVSL value of $9,576 (assuming a 2% discount rate). This is not only below the assumed morally reasonable threshold for saving a life but is also below the net yielded value of buying the body armor. Therefore proportionality does not require buying the shielded munitions over the body armor – even though more lives would thereby be saved. But there does not seem to be anything morally relevant about the fact that this case is framed in terms of expenditure decisions. We can imagine a functionally equivalent case in which a commander can alter their conduct of operations in such a way as to either protect 100 persons now or 110 persons in 10 years. If the future loss of life should be discounted for the proportionality case involving expenditure, then there does not seem to be any reason not to discount in nonexpenditure cases.

There is a final issue we must consider about the account I have given of value discounting over time. The fundamental reason that monetary values are discounted over time is that monetary capital is *productive*. When I conclude that it is better to have $10,000 today rather than $11,000 in ten years, this is because in the interim period the $10,000 can be invested to produce a 2 percent annual rate of return. Economists typically assume that people will always act in an economically rational – that is to say maximizing – way. Therefore, they assume that money will be invested at the best available risk-adjusted rate of return. But clearly, this does not always happen in the real world. For the purposes of moral assessment, what matters for the discount rate is not the theoretical available rate of investment return, but the probable *actual* rate of return. If you are just going to put the $10,000 under your mattress, then you are better off choosing the option of $11,000 in ten years' time. This means that for the purposes of determining how much we can reasonably be required to spend today to save a future life, the discount rate we should use will depend on the likely *actual* use of those funds assuming they are not deployed to save a life. Will they be invested productively at a rate of 2 percent or 5 percent or 10 percent? Or will they be frittered away on frivolous consumption? Since many funds will not in fact be invested, but simply diverted to consumption, it is likely that the actual discount rate should in most cases be substantially lower than available market rates of return.

What is more, I have argued that moral harms and goods should also be discounted over time. If we are to apply the discounting logic to moral, as well as monetary goods, then it must be the case that some moral goods, like money, can similarly be productive over time. If this were not the case, then we would have no reason to prefer a present moral good to a future moral good (just as we have no reason to prefer $100 today over $100 next year if we are just going to put the money under a mattress).

What does it mean for a moral good to be productive? A moral good is productive if it has the capacity to generate additional moral value over time. A human life is clearly morally productive in this way. If I save your life, then I create (or secure) a moral good for you. But in addition to being a locus of moral value in itself, your life creates a ripple of ancillary goods through time. It enables ongoing love and companionship for your friends and family, productivity for your employer, education for your students, tax revenue for the state. If you choose to have children, it produces the potential for a chain of moral value creation stretching indefinitely into the future. The opportunity costs created by this moral fecundity are the underlying reason that future moral goods are discounted compared with present goods (just as the opportunity costs created by the potential for investment are the underlying reason for discounting money). When I save a life today, compared to saving a life in ten years' time, I secure an additional ten years of the production and enjoyment of ancillary moral goods which would otherwise be lost forever.

In contrast to the value of a life, many consumption goods are not productive in this way. If I spend money on a fine meal, my pleasure and enjoyment are moral goods. But it generates little in additional moral value over time. However, if I commission a great work of art, or plant a garden, this can generate enduring moral value far into the future by creating opportunities for reflection, repose, and aesthetic enjoyment. So, the distinction between productive and nonproductive moral goods is not coextensive with the economist's distinction between (productive) investment and (nonproductive) consumption of financial resources.[8]

There is a second important difference between the ways in which moral goods and money can be productive. When money is invested, the interest is most often compounded, generating an exponential rate of growth (the present value formula assumes compound interest to calculate the relationship between present and future value). Moreover, the returns from financial investment are assumed to continue indefinitely into the future. It is the exponential and indefinite rate of growth that creates the extraordinary and grotesque sounding disparities between present and future value that so worry the critics of social discounting (for example Parfit's observation that, assuming a 5% discount rate, one life today is worth more than a billion lives 400 years from now).

But productive moral goods possess neither of these features. A human life creates additional moral goods through time, as we have noted. But the production of those goods does not increase exponentially, and they

[8] We have discussed how moral goods can be productive by generating further goods through time. But goods can also, of course, be negatively productive, generating moral harms through time. One example would be if the life you save turns out to belong a pathological murderer. Another example is the environmental damage caused by yourself, your children, and their descendants. For these reasons it is possible that in certain circumstances that the social discount rate may be negative, implying that future goods have a higher present value than present goods.

do not continue indefinitely. Rather they wax and wane over the course of a life. Moreover (with the exception of having children) most moral goods do not create additional value indefinitely. Our capacity to work productively and to give love and affection wanes with age and then ceases with death. Cultural artifacts eventually fall into dust and cease to provide pleasure and edification. For these reasons, the present value of future moral goods will need to be calculated in a very different way to that represented in the present value formula. The discount rate will likely need to be lower than that commonly assumed in calculating the present value of financial assets. The discount rate will be linear and most likely variable rather than exponential, and discounting will apply over a limited period of time only rather than extending indefinitely into the future. Once these differences are properly taken into account, I do not believe that the application of an appropriate social discount rate to future moral goods and harms will yield any implausible or morally objectionable implications.

Conclusion

In this chapter, I have used the idea of endless wars to explore the impact of time on the ethics of war. When wars extend over long periods of time and beyond the regular planning cycle of leaders and policymakers, a raft of ethical issues are raised. While it is not impossible that such wars meet the proportionality and reasonable prospect of success conditions, it is unlikely they will do so, and moreover, such assessments suffer from significant indeterminacy problems. A significant ethical implication of endless wars concerns the effect of time on *ad bellum* proportionality judgments. The discounting of the present value of future costs and benefits plays an important role in many policy areas including optimal investment theory and climate change, but it has not typically been applied to the ethics of war. But endless wars force us to revisit this.

I argue that we need to define a measure for the "moral value of a statistical life" in order to resolve problems concerning the reasonable expenditure required to save a life and that the present value of this measure must be discounted over time if we are to preserve a principle of intertemporal neutrality. I then extend this reasoning to wide proportionality more broadly and provide an account of the underlying moral reasons for the discounting of moral values through time rooted in the idea of productive moral goods. The issue of discounting in the ethics of war merits much greater investigation, and I hope this discussion will have helpfully framed some of the key points for future discussion.

CHAPTER 8

Two Conceptions of the Proportionality Budget for Jus Ex Bello

Darrel Moellendorf[*†]

Wars drag on. They can continue long beyond their forecasted end. Whether this is due to wishful thinking on the part of leaders at the onset of the war, culpable ignorance about the prospects for victory, impossible to foresee happenstance, or some combination of all of these, the fact remains that wars too often go on much longer than participants or supporters expected at the onset. There is no better recent example of this than the war that began with the U.S. invasion of Afghanistan in response to the terrorist attacks of September 11, 2001. It was a full two decades before the U.S. military and NATO allies pulled out and the Taliban returned to power after apparently being defeated very early on. In fact, there were good reasons to doubt the prospects for success even from the beginning, although these were largely ignored.[1] The condition of the likelihood of success is related to the constraints of proportionality. A war that can only be won by imposing disproportional damage is one that cannot succeed within moral constraints. Success is limited by proportionality. This is certainly relevant to the long war in Afghanistan. Even if decisive victory proved possible, it is of no matter if the costs of success are disproportional.

So much depends on what and how much may be imposed in pursuit of a just cause. Pursuit of a just cause may not proceed by all manner of death and destruction.[2] On the contrary, regardless of the justice of the cause, a war may be unjust to initiate because its pursuit could only be successful by imposing excessive harms. The concept of proportionality holds that there is a limit on the harms that may be imposed in the pursuit of the right or the good.

* I would like to thank Lukas Sparenborg for his comments on the penultimate draft of this paper and for his help with the formatting.

1 See Darrel Moellendorf, "Is the War in Afghanistan Just?" *Imprints A Journal of Analytical Socialism*, 6, no. 2 (2002).

2 In contrast, Michael Walzer holds that considerations of proportionality play only a very minor role in evaluating the justice of a war. See Michael Walzer, *Arguing about War* (New Haven, CT: Yale University Press, 2004), p. 91.

I call the set of maximum acceptable harms that might be imposed in the pursuit of the just cause *the proportionality budget* of the war. In *jus ad bellum* the proportionality budget is compared to the best empirical forecast of the reasonably likely harms of the war in order to determine whether initiating a war would be consistent with the proportionality constraint.

Suppose the few skeptics were wrong about the U.S. war in Afghanistan; suppose that at its initiation it was reasonably likely to succeed within its proportionality budget. But later, say more than a decade later, it had not succeeded; and it had exhausted its proportionality budget. What then? May it be continued?[3] Just as there are constraints on initiating a war, there are constraints on the continued prosecution of a war. These constraints comprise one of the doctrines of what I call *jus ex bello*.[4] There are two major categories of concern for *jus ex bello*: First, should the war be continued or terminated? Second, if terminated, how?[5] With regard to the first concern, considerations of proportionality would seem to be as important as they were for *jus ad bellum*. The same reasons for constraining the pursuit of just cause by some maximum of harms imposed would seem to apply. No aim, however just, could be pursued if it required limitless harm. In considering whether to go on pursuing a just cause by means of war, what significance should the exhaustion of the war's correctly determined initial proportionality budget have in determining whether it may be continued?

Discussion of this question has distinguished two broad views.[6] One view holds that the fact of budget exhaustion matters decisively; insofar as continued prosecution would impose additional harms, it would be unjust because it is disproportional. The second argues that the exhaustion of proportionality budget matters not at all; the only considerations are prospective. Moving forward all that matters is the reasonably foreseeable harms and

[3] See Darrel Moellendorf, "*Jus ex Bello* in Afghanistan," *Ethics and International Affairs*, 25, no. 2 (2011), 155–64. See also Richard W. Miller, "The Ethics of America's Afghan War," *Ethics and International Affairs*, 25, no. 2 (2011), 103–31; and Jeff McMahan, "Proportionality in the Afghan War," *Ethics and International Affairs*, 25, no. 2 (2011), 143–54.

[4] See Darrel Moellendorf, "Jus ex Bello," *Journal of Political Philosophy*, 16, no. 2 (2008), 123–36.

[5] See Darrel Moellendorf, "Two Doctrines of *Jus ex Bello*," *Ethics*, 125, no. 3 (2015), 653–73.

[6] Representative examples of the view that proportionality budget exhaustion entails that continuing the war would be unjust are Cécile Fabre, "War Exit," *Ethics*, 125, no. 3 (2015), 631–52, Moellendorf, "Two Doctrines of *Jus ex Bello*." Representative examples of the view that only proportionality going forwards matters are Jeff McMahan, "Proportionality and Time," *Ethics*, 125, no. 3 (2015), 696–71 and David Rodin, "The War Trap: Dilemmas of *jus terminatio*," *Ethics*, 125, no. 3 (2015), 674–95. A middle path would be that the past matters, but as much as the future due to discounting. This is discussed in Seth Lazar, "Moral Sunk Costs," *The Philosophical Quarterly*, 68, no. 273 (2018), 841–61.

whether they are proportional to the aim considered anew. I have supported the first view, contending *inter alia* that the position of the second camp is untenable because, in principle, a war may be prosecuted indefinitely if after reconsidering the proportionality budget each time that its limit has been reached prospectively the war appears proportional. The second camp argues among other things that it would be irrational not to license an aim that can be pursued within acceptable prospective proportionality budget.

The concept of proportionality implies some kind of equilibrium point between the value pursued and the disvalue created. Conceptions of proportionality differ according to how that equilibrium point is understood and justified.[7] Various conceptions of proportionality yield different means of calculating the proportionality budget and different understandings of what is at stake when the budget is exceeded. In this chapter, I work through two different conceptions of proportionality, and I ask whether these make a difference to the debate within *jus ex bello* about continuing a war that has reached the limit of its *ante bellum* proportionality budget. I call these the *Impersonal Value Conception* and the *Personal Value Conception* of proportionality. I argue that the Personal Value Conception of proportionality has grounds to be more restrictive of the harms that may be imposed in the pursuit of a just cause.

The Impersonal Value Conception

The Impersonal Value Conception expresses the ideal of equilibrium in commensurable impersonal value and disvalue. The conception weighs the disvalue of the costs of successfully prosecuting the war against the value of securing the just cause. In order to judge whether the requirements of *jus ad bellum* are satisfied, one must aggregate the foreseeable costs and weigh them against the better condition realized by success.[8] If the costs do not outweigh the better condition realized, the war is proportional.

Let's abstract from war to consider a formal case in which the Impersonal Value Conception arises in the assessment of costs in the pursuit of a better condition. I shall call it *Base*, and it consists of at least two points in time, *Ante* and In_1:

[7] An informative discussion of different conceptions of proportionality is to be found in Thomas Hurka, "Proportionality and the Morality of War," *Philosophy and Public Affairs*, 33, no. 1 (2004), 34–66.

[8] As Francis Kamm puts it, "Stopping aggression may be justify starting a war only if the direct bad effects of a war are proportional to the good effects that will result." Francis Kamm, *Ethics for Enemies* (Oxford: Oxford University Press, 2011), p. 131.

Ante: There is a possible better condition of great value, V. Remaining neutral about whether or not its value is deontic, we are *pro tanto* either permitted or required to pursue the condition within costs. The limit of the proportionality budget for realizing V is the equilibrium point beyond which the aggregate costs of pursuing the better condition would exceed the value gained. There is sound reason, using all the relevant available evidence, to believe that the condition is achievable at aggregate costs $-C$. C can be compared to V, and $C \leq V$.

In_I: At some point in the pursuit of the valuable condition, before its realization, quite unexpectedly the proportionality budget has been expended. The successful realization of the better condition of value V is once again with sound reason, using all the relevant available evidence, estimated to be achievable at costs $-C_2$, and $C_2 \leq V$. Although taking into account the costs of the entire effort, it could cost much more than the original proportionality budget, the future costs alone of the project are reasonably expected to be within budget.

The reasoning at point In_I is epistemically constrained at least insofar as it cannot rule out the possibility of subsequent points in time that instantiate exactly the same moral considerations. Although the available evidence makes it unlikely, there could be an instance *In2* in which the better condition, although still not realized, could be realized at cost $-C_3$ and $C_3 \leq V$. Any time at which the relevant circumstances occur is then an instance of In_X. Consider an argument that might be made at In_I to continue to pursue V or to halt the pursuit.

Continue Base: The argument for continuing is identical to the argument at *Ante*. The positive value of the foreseeable costs of achieving the better condition is less than or equal to the value of the condition. *Ceteris paribus*. Continued pursuit of the better condition is either permitted or required by the Impersonal Value Conception in comparison to the foreseeable costs.

Now consider a variation of *Base* that I shall call *Variant*. *Ante* is the same, but there is a difference in the kind of better condition realized that makes for a difference at the second point in time:

In_1 Variant: At some point in the pursuit of the better condition, but before its realization, unexpectedly the proportionality budget has been expended. The better condition pursued is partially realizable, and its value V is aggregative. The better condition has been partially realized. The value realized so far is V_1. There is a portion of value remaining to be realized < V. This remainder is V_2; and $V_1 + V_2 = V$. The successful realization of the rest of the better condition would yield additional value V_2, which is with sound reason, using all the relevant available evidence, estimated to be achievable at costs $-C_2$, and $C_2 \leq V_2$.

The reasoning at *In_1 Variant* is also epistemically constrained as above. So, again any time at which the above circumstances occur is an instance of *In_X Variant*. Two arguments can be made at *In_1 Variant* for continuing the pursuit of the full realization of the valuable condition.

$Continue_1$ Variant: The positive value of the foreseeable costs of achieving the remaining value is less than or equal to the value of the remaining condition to be realized. *Ceteris paribus*, depending upon the deontic value of the condition, continued pursuit of the better condition is either permitted or required by the Impersonal Value Conception given the foreseeable costs.

$Continue_2$ Variant: $V_1 - C$ is the net (negative) value at time *In_1 variant* imposed by efforts to achieve the valuable state of affairs. $(V_1 + V_2) - (C + C_2)$ is the net foreseeable (negative) value of achieving the valuable state of affairs. If the latter is greater than the former, then continuation would be the least costly measure. The latter would be greater than the former as long as $V_2 > C_2$. In other words, imposing additional costs can reduce net costs if the better condition realized is sufficiently high, in which case not continuing is net more costly. Now if the equilibrium between costs and the value of the better condition realized cannot be achieved, then surely net costs should be reduced. Pursuit of the least costly measure in excess of the proportionality budget amounts to seeking the least disproportional outcome. According to the method of calculation used in this conception, it is possible for an effort that exceeds the original budget to be less net costly than one that is constrained by the original budget.

Let's consider arguments that might be made at the second point in time against continuing the pursuit of the better condition in excess of the original proportionality budget.

> *Halt$_1$ Base:* The value of the better condition is *V*. To continue to pursue the better condition would be to incur costs that exceed that value. The limit to cost imposition as understood by the Impersonal Value Conception of proportionality is the equilibrium between costs and value obtained. Only halting satisfies that equilibrium. Halting is therefore required by the Impersonal Value Conception.
>
> *Halt$_2$ Base:* The logic of *Continue Base* allows in principle the unending imposition of costs. Just consider the possibilities of *In_X* and *In_X Variant.* The concept of proportionality is that there is a limit on the pursuit of all value. By licensing unending harm imposition, *Continue Base* is not only inconsistent with the Impersonal Value Conception but also with the very concept of proportionality.

The point of both lines of argumentation is to focus on the whole effort to realize the better condition. The two *Continue* arguments seem plausible only if the harms assumed at *In_X* or *In_X Variant* are fully discounted in the accounting of the harms imposed to achieve the better condition. For if they were included, the judgment would have to be that to continue is disproportional.

The importance of the distinction between *Base* and *Variant* is that the reasons to continue at *In_X variant* are stronger than those at *In_X* insofar as the former take into consideration not only forward-looking considerations, but consideration of the net value of the entire effort. Although making the case for continuing in *Base* requires discounting past costs, the argument of *Continue$_2$ Variant* responds to *Halt$_1$ Base* and *Halt$_2$ Base* on their own terms by taking into consideration past costs. *Continue$_2$ Variant* counts the disvalue of past costs at their full value, and in doing so enables the following plausible line or response: To halt the pursuit of the better condition would be irrational since the costs that are the *pro tanto* reason to halt are reasonably expected to be net lower by continuing. Taking the comparative weight of costs seriously, as the Impersonal Value Conception is committed to, permits or requires continuing the pursuit of the better condition.

In light of the force of *Continue$_2$ Variant,* perhaps the strongest argument against continuing is the following:

Halt₃ Base: The cost calculation that results in C_2 takes into effect only the costs directly caused by the pursuit of the better condition. Any adequate reckoning of the costs of continued pursuit of the better condition has to consider the indirect costs, especially the reasonably foreseeable opportunity costs, the costs of the options foreclosed by pursuing the better condition. When these are added, the sum C_2^+ could exceed the value, V_2. If that is the case, then the conclusion of *Continue₂ Variant* does not follow.

Two things about this third argument are noteworthy. First, it would seem *ad hoc* unless the original proportionality budget also includes the foreseeable opportunity costs of the pursuit of the better condition. If, in general, one must include foreseeable opportunity costs in the consideration of the morality of pursuing a better condition or acting on a duty, then the case for permissible or required action is more constrained, including at the beginning of the enterprise. Second, the force of the response is contingent rather than categorical. It accepts the logic of *Continue₂ Variant*, but the set of cases in which there would be permission or obligation to continue would be smaller, although not necessarily null.

According to one defense of the Impersonal Value Conception of proportionality, proportionality requires that our projects in pursuit of value contain costs according to the ideal of least net costs. The constraint of least net costs provides a response to the worry that this conception permits unlimited costs in violation of the concept of proportionality. Every additional cost is justified just insofar as it reduces net costs. Importantly, according to this view, the initial proportionality budget plays a role. After costs have reached the budget limit, they are justified only if they reduce net costs.

Let's consider a realistic example in which the form of the debate sketched above might play out.

Art Restoration: Valuable works of visual art whose properties have become dulled over the years from exposure are held in a public art gallery. Their restoration would be of great aesthetic value. The proportionality budget is the market value equivalent of the aesthetic value, however calculated. This is used to consider whether the costs of restoration can be assumed. The costs of restoration include the market value of the labor, the know-how, the materials and equipment, and the opportunity costs of forgoing other public expenditures. Insofar as the reasonably foreseeable costs are less than or equal to the added value of the restoration, there is an equilibrium that allows the pursuit of the project of restoration.

Suppose that the initial allowable budget for the restoration has been spent, and, due to unforeseeable difficulties, the art has not yet been fully restored. Now there is the reasonable expectation that it can be restored at greater cost, and the forecasted costs taken alone would not exceed the value added of the restored art. Should the full costs of restoration exceed the value of the product? With *Base* and *Variant* in mind, we imagine that one camp argues that stopping now would only incur greater loss, whereas restoring the art further would bring significant aesthetic value and importantly less net loss. In contrast, the other camp argues that the pursuit of restoring the valuable items must stop because it is irrational to exceed the value of the restoration; moreover, budgets are tight, and the opportunity costs of expenditure overruns are high.

One can imagine facing difficult issues of commensurability and doubts about whether the art is correctly valued either initially or on the second occasion. The latter problem we assume is taken care of by a *correct* evaluation of the market value of the art. This allows us to focus on whether the original budget should be overrun. The former question of commensurability haunts all cases of the Impersonal Value Conception, and perhaps it is a weakness worth noting, but the accounts simply assume there is some kind of practical resolution to it. The practical answer in this case would probably include noting that one simply must judge how much money one should spend on restoration. Setting a budget requires assuming some kind of commensurability. In principle, rational disagreement about that might exist insofar as the values are incommensurable, but once some measure of commensurability is set for the initial calculation of the proportionality budget, practical rationality requires that it be adhered to where possible when considering whether or not to act in a way that would involve cost overruns.

Whether or not restoration should be continued once the restoration budget has been exhausted would seem to depend upon the question of the opportunity costs. The opportunity costs of the initial budget may have been proportional, but the public expenditures that would be foregone by additional expenditures may be too much. In any case, if someone were to oppose continuing the expenditures on restoration on grounds of opening the door to assuming infinite costs, that would fail to appreciate that the justification of continued expenditure is to pursue the lowest net costs.

Keeping *Art Restoration* in mind, it is not hard to imagine a war case that, although vague on details, is clear enough to set out how the concept works and where the controversies are.

> *Injustice Remediation*: A regime has perpetrated serious injustices and continues to maintain the conditions of injustice, continues in at least that sense, but perhaps also more actively, to perpetuate injustices. We leave the idea vague now, so that in principle it could cover various, different conceptions of just cause. The evidence of feasible nonlethal alternatives for successfully achieving remediation is little at best. Alternatively, a belligerent response seems reasonably likely to succeed (whatever the appropriate degree of credence is). Additionally, the value of remediation is sufficiently high that, given the available evidence, the reasonably foreseeable direct and indirect costs are not excessive. The proportionality budget limits those costs by value of the injustice remediated. The pursuit of the remediation of the injustice reaches the budget limit, but the injustice has not been completely eliminated. Some forms of grave injustice continue. There is still far too little prospect of peaceful remediation, and sufficient evidence of the effectiveness of war. Moreover, the available evidence is strong that the positive value of the direct costs of war will not exceed the value realized; $V_2 > C_2$.

It would be plausible to object to continuing the war if there are reasons to believe that the opportunity costs were sufficiently high. For example, when Martin Luther King, Jr. objected to continuing the Vietnam War he did so because, among other things, it took away from the U.S. government's effort to eradicate poverty.[9] If we assume, for the sake of this example, that the positive value of the combined reasonable, foreseeable, direct and indirect costs are less than the value achieved, then the continued prosecution of the war satisfies this conception of proportionality. It would not seem to be relevant to claim that accepting this principle opens the door to unlimited costs since reasonably expected net costs are at their highest at this point. Continued prosecution of the war is justified only if it reduces net costs. Nor can one object at this stage that the unintentional killing of nonliable people is incommensurate with value added by the

[9] See Martin Luther King Jr., *A Testament of Hope: The Essential Writings of and Speeches of Martin Luther King, Jr.*, James M. Washington (ed.) (New York: Harper Collins, 1991), pp. 233, 635.

war's continued prosecution. Although that worry raises important philosophical questions of commensurability, the initial justification of the war included the requirement that it satisfy proportionality; therefore, the example assumes commensurability. Under these conditions, continuing the war would seem to be proportional.

The Personal Value Conception

The characteristic feature of the Impersonal Value Conception is that proportionality guides the pursuit of projects to ensure a maximal ratio of good outcomes to costs imposed. It is then an optimizing conception of proportionality based on a conception of value that is, following Thomas Scanlon, teleological.[10] If, and because, an object is valuable, one has reason to seek to promote it.[11] Due to the distinction he draws between the dignity of persons and the price of things, Immanuel Kant is often acknowledged as the fount of thinking that objects to the teleological conception of value.[12] But Kant was following Samuel Pufendorf, who claimed that "In the very name of man a certain dignity is felt to lie, so that ultimate and most effective rebuttal of insolence and insults from other is, 'Look I am not a dog, but a man, as well as yourself.'"[13] The harm that Pufendorf imagines one objecting to is not a failure to promote some value, but instead a failure to respect a person. The rejection of the teleological conception of value denies that all value has the character of that which is to be promoted.[14] Human beings, specifically, are not valuable in that way. And insofar as that is the case, human morality and justice are not captured by a normative theory employing a teleological conception of value. One idea is that respecting persons

10 Thomas Scanlon, *What We Owe to Each Other* (Cambridge, MA: Harvard University Press, 1998), pp. 79–87.

11 Ibid. See also Derek Parfit, *On What Matters* (Oxford: Oxford University Press, 2011), vol. 1, p. 236.

12 See Immanuel Kant, *Ethical Philosophy*, second edition, (Indianapolis, IN: Hackett Publishing, 1995), *Grounding for the Metaphysics of Morals*, (Ak. 434–435), p. 40. One prominent example of invoking Kant this way can be found in Parfit, *On What Matters*, vol. 1, pp. 239–44.

13 Samuel Pufendorf, *On the Duty of Man and Citizen According to Natural Law*, James Tully (ed.) (Cambridge: Cambridge University Press, 1991), p. 61 (bk. I, chp. 7, sec. 1).

14 I am not claiming that this is the only plausible way to read Kant and Pufendorf. One might claim that it makes more sense to defend a pluralistic world of impersonal values, all of which share the characteristic of that which is to be promoted, in which some values are morally more important than others. Thomas E. Hill mentions but disregards such a view in his "Making Exceptions Without Abandoning the Principle: Or How a Kantian Might Think About Terrorism," in Raymond Gillespie Frey and Christopher Armstrong (eds.), *Violence, Terrorism, and Justice* (New York: Cambridge University Press, 1991), 204–5.

requires appropriate responses both to their value as persons worthy of respect and to the value of various human goods.[15]

This suggests the possibility of an alternative conception of proportionality that takes the limit of the budget to consist in something other than the promotion of the optimal ratio of impersonal value to disvalue. The Impersonal Value Conception might be perfectly appropriate in cases where our action concerns impersonal value, cases such as in *Art Restoration*, and yet not in cases in which our relationships to other human beings are at stake.

Clarifying the Personal Value Conception requires specifying the nature of the equilibrium point between the value of the end achieved or of the duty pursued and the disvalue of the damages or wrongs of acting. Let's begin with the idea that we value human beings appropriately when we respect them, and additionally that respecting them requires treating them, unlike things, such as works of art, in ways that could be justified to them.[16] The equilibrium point between a valuable or required end and the damages incurred, which are *pro tanto* wrong to persons, is defined by a principle, *P*, which could be justified to others. Justification of *P*, let's suppose, involves some kind of moral contractualism. For purposes of illustration, one could imagine, following Scanlon, that the equilibrium point is defined by a principle that no one would have reason to reject.[17]

Exploring the Personal Value Conception of responsibility requires modifying the earlier examples. I start with what I call *Modified Base*.

> *Ante*: There is a possible more just condition, *J*. We are *pro tanto* either permitted or required to pursue the condition within moral limits. The relevant limits define permissions to commit *pro tanto* wrongs to persons while pursuing *J*. There is sound reason, using all the relevant available evidence, to believe that the condition is achievable, doing so will involve foreseeably committing *pro tanto* wrongs - *W*. There is a principle, *P*, according to which it would be

[15] Scanlon, *What We Owe to Each Other*, p. 172.

[16] There are various important statements of this view. For some of them see Scanlon, *What We Owe to Each Other*, p. 106; Parfit, *On What Matters*, vol. 1, p. 239; Thomas E. Hill, "Making Exceptions," 214, Stephen Darwall, *The Second Person Standpoint* (Cambridge, MA: Harvard University Press, 2009), pp. 300–20, and Rainer Forst, *The Right to Justification* (New York: Columbia University Press, 2011), pp. 18–23.

[17] Scanlon, *What We Owe to Each Other*, p. 153. The general points here about the Personal Value Conception, however, are not tied narrowly to Scanlon's criterion of justification.

permissible to pursue *J* up to $-W$; and that principle is justified to those subjected to it.

In_1: At some point in the pursuit of the more just condition, before its realization and unexpectedly, wrongs $-W$ have been committed. The proportionality budget is at the limit. However, the successful realization of the better condition *J*, once again with sound reason and using all the relevant available evidence, is estimated to be achievable but would involve committing wrongs $-W_2$, where $-W_2 \leq -W$.

As before, consider an argument that might be proffered at In_1 to continue to pursue *J* or to halt its pursuit.

Continue Modified Base: The argument for continuing is at least as strong as the argument at *Ante*. If according to *P* it is permissible to pursue *J* up to $-W$, then insofar as $-W_2 \leq -W$, it must be the case that $-W_2$ is permitted by *P*. Hence, the Personal Value Conception permits continuing the effort in excess of the original proportionality budget for *pro tanto* wrongs.

By way of contrast, consider two arguments against continuing the pursuit of greater justice in excess of the original proportionality budget.

$Halt_1$ Modified Base: To continue to pursue the better condition would be to commit wrongs $-W$ and $-W_2$, which are more than *P* allows. The limit to the *pro tanto* wrongs that may be imposed, as understood by the Personal Value Conception, is the equilibrium as defined by a principle that can be justified to others. Halting is necessary to ensure that the wrongs do not exceed the equilibrium point. Therefore, the Personal Value Conception requires halting.

$Halt_2$ Modified Base: The method of *Continue Modified Base*, which is to count only forward-looking *pro tanto* wrongs, allows, in principle, for the unending imposition of wrongs anytime prospective wrongs are less than the limits of the original principle. But the concept of proportionality is that there is some limit to the *pro tanto* wrongdoing that may be done in pursuit of justice. *Continue Modified Base* is, then, not only inconsistent with the Personal Value Conception but also with the very concept of proportionality.

The Personal Value Conception is not based on the ideal of promoting aggregate impersonal value. Instead, it incorporates the idea of respecting persons, which it takes as requiring action in accordance with principles that can be justified to others. It is an open question as to which considerations the justification of a principle to others would require. The principle could theoretically include several considerations in addition to positive outcomes, including, *inter alia*, their distribution, their causal origin, their importance, and agential intent. What in fact gets taken on board depends on what it would be reasonable for parties to consider.

In contrast to the Impersonal Value Conception, there is good reason to believe that for the Personal Value Conception *P* must include a clause that requires the full counting of all *pro tanto* wrongs, past as well as prospective, created by the pursuit of greater justice. An important reason in favor of the full counting clause is stated in *Halt$_2$ Modified Base*, namely that not counting past wrongs allows never-ending instances of wrongdoing, which seems incompatible with the very concept of proportionality. Absent counting past as well as prospective *pro tanto* wrongdoing, there is no end to wrongdoing permitted.

Now according to the Impersonal Value Conception there is a reply to this worry of unending cost imposition. In *Continue$_2$ Variant*, continuation is allowed insofar as and because it improves the ratio of benefits to costs. This seems unavailable to the Personal Value Conception. Suppose that *P* contained a clause that held that *pro tanto* wrongs may continue beyond the original limit just insofar as and because the good achieved by doing so reduces the net costs. Persons who would suffer the additional wrongs could object that the relationship between wrongdoing and justice simply is not such that the latter can make up for the former.

The wrongs they would suffer are not in any way redeemed by the greater justice achieved. The Impersonal Value Conception allows this only because of assumptions, which the Personal Value Conception does not make. I explore two of these assumptions in the next section.

Distinctions in Value

The Impersonal Value Conception makes two assumptions that are not available to the Personal Value Conception, but these assumptions are necessary in order to justify the continued prosecution of a war that has reached the limit of its proportionality budget. One assumption can be understood by distinguishing between smooth and chunky types of justifications. Justification of the proportionality budget according to the

Impersonal Value Conception is smooth in the sense that the justification closely tracks marginal differences in the ratio of benefits to costs *and* the ratio can increase incrementally as actions yield marginal benefits. If the ratio of benefits to costs increases as additional costs are imposed, an objection to the war on grounds of proportionality seems to miss the point. In contrast, a form of justification not sensitive to marginal improvements or not *only* sensitive to them is less able to avail itself of this sort of response to the objection that the imposition of disvalue is endless.

Nonteleological conceptions of value and nonconsequentialist moral theories reject justification solely on the basis of the optimal ratio of benefits to costs. For example, Rawls's difference principle states a threshold that Pareto improvements in wellbeing must satisfy. Robert Nozick argues that rights serve as side constraints on the pursuit of optimization.[18] And Derek Parfit has observed that the contractualist ideal of consent offers a veto to persons against being made to share the burdens imposed by optimization.[19] The Personal Value Conception opens up the possibility that events must satisfy a threshold other than or in addition to optimization in order to be justified, or that events may lose their positive valence when they contravene side constraints. In this way, the Personal Value Conception employs a chunky type of justification.

Chunky justifications cannot confidently allow the imposition of additional disvalue on grounds that it can be offset by the greater benefits. If *P* is justified if and only if it is acceptable to others, and if the additional disvalue would include violations of human rights, there is no assurance that *P* would take such *pro tanto* wrongs as in equilibrium with the benefits. Additionally, there is no guarantee that the pursuit of greater overall net benefit is justified if there are contractualist constraints on the distribution of the benefits. The chunkiness of the justification of the proportionality budget, according to the Personal Value Conception, casts doubt on a principle that allowed iterative permission to continue to pursue the better outcome.

A second important distinction is between imposing costs and disrespecting persons. The Impersonal Value Conception takes disvalue as costs imposed, which are outweighed by benefits. In that way, one can sensibly talk about net gains. In *Art Restoration* consideration of the trade-off between the market values of the result and the labor and equipment costs

[18] Robert Nozick, *Anarchy, State, and Utopia* (New York: Basic Books, 1974), pp. 28–33.

[19] Parfit, *On What Matters*, p. 356.

is a perspicuous example of that kind of accounting. That there must be an equivalent that lends coherence to such moral mathematics is made clear by such an example, which employs the purported universal equivalent, money. Monetary costs can be redeemed by subsequent gains, at least if the opportunity costs of continuing can be borne. Theodor W. Adorno once complained that the spread of the exchange principle imposes a kind of false identity on the whole world, the fiction of a universal equivalent in value.[20] The Personal Value Conception rejects the assumption of a universal equivalent in value. Value is fragmented in a manner that includes the requirement of responding to persons with respect rather than with efforts of promotion.[21]

Failures of respect are forms of wrongdoing that are not reduced by the production of greater net impersonal good. Costs can be aggregated and compared to benefits in a way that permits drawing up a balance sheet of net gain. And to the extent that this can be done across time, costs imposed at time t_n can be offset against benefits yielded at time t_{n+1}, such that the subsequent benefit enjoyed can redeem the costs created. Foregone consumption invested wisely can generate greater income enjoyment later. Acts of disrespect for the sake of greater justice are troublesome for attempts at redemptive calculation since a basis of comparison and a capacity for aggregation are doubtful. This renders doubtful the idea that committing acts that are disrespectful can be redeemed if they were to yield a net better condition.

These two assumptions make it possible for the Impersonal Value Conception of proportionality to permit the continued prosecution of a war that has exceeded its proportionality budget if *ceteris paribus* continuing is reasonably expected to yield a higher ratio of benefits to costs. The Personal Value Conception abjures these assumptions, and therefore lacks the bases upon which to endorse continuing in disregard of the original proportional budget.

Disrespect and the Pacifist Challenge

Proportionality in just war theory constrains permission to kill and destroy. I have argued that the Personal Value Conception offers no reasons to permit the commission of wrongdoing beyond that allowed by the original

[20] See Theodor W. Adorno, *Negative Dialectics* (New York: Continuum Press, 1981), p. 146.

[21] On the fragmentation of value see Thomas Nagel, *Mortal Questions* (New York: Cambridge University Press, 1979), pp. 128–41.

proportionality budget, even if doing so would promote greater justice. This, I have argued, is because it does not allow trading acts of disrespect against the creation of better conditions. But if trading acts of disrespect against the creation of better conditions is the problem, how can the Personal Value Conception permit killing any people, even unintentionally, in the pursuit of a just cause? One influential line of pacifist criticism is that the foreseeable killing of innocents, albeit unintentionally, is wrong, and such a wrong is unavoidable in conditions of modern warfare.[22] The response of the Impersonal Value Conception is to weigh the disvalue of the lives lost by killing against the good created by prosecuting the war. The previous two suggestions have argued that such an approach is not available to the Personal Value Conception of proportionality. Why wouldn't the proportionality budget for the Personal Value Conception be zero?

As Jeff McMahan explains it, proportionality governs the following four distinct kinds of actions: Intentional harm to the potentially liable; unintentional but foreseeable harm to the potentially liable; intentional harm to the not potentially liable; and unintentional but foreseeable harm to the not potentially liable.[23] Wide proportionality governs the latter class of actions. This kind of proportionality generally includes concerns about the killing of noncombatants, which is unavoidable in modern forms of war. Such killing is surely *pro tanto* disrespectful and wrong, typically a great wrong. If such harm cannot be redeemed by subtracting its disvalue from the value achieved, how can the wrong be justified? Because the Personal Value Conception rests on the idea of respect for persons, its capacity to account for judgments of wide proportionality in a way that would rebuff the pacifist challenge might be in doubt.

Some kind of lesser evil argument would be required, but it could not be one that merely weighs impersonal disvalue against impersonal value. A principle of proportionality regarding foreseeable but unintentional harms imposed on those not liable to them, according to the Personal Value Conception of proportionality, would hold that the imposition of such harms is permissible to the extent that there is a P that licenses such killing, and that P is not met with reasonable objections. A principle of optimal impersonal value promotion is objectionable on grounds

[22] See for example Robert L. Holmes, *On War and Morality* (Princeton, NJ: Princeton University Press, 1989), chap. 6. Holmes explicitly rejects appeals to the lesser evil of the achievement of the just cause: "The killing of innocents by an aggressor is no worse *as such* that the killing by those who would oppose him by waging war"; 211. See also the discussion in Richard Norman, *Ethics Killing and War* (Cambridge: Cambridge University Press, 1995), chaps. 5–6.

[23] Jeff McMahan, *Killing in War* (Oxford: Oxford University Press, 2009), pp. 19–20.

that it uses persons merely as means for the greater value. In his discussion of a broadly Kantian approach to the permissibility of a public policy permitting foreseeably killing innocents in an effort to respond to terrorism, Thomas Hill Jr.'s comments seem to suggest the following principle:

> A policy of responding to terrorism in a manner that includes limited foreseeable but unintentional killing of innocents is permissible only if each person (citizen) has a compelling reason to accept the risks of being killed.[24]

The idea is that although there are burdens to the acceptance of such a principle, namely being permissibly killed by law enforcement, each person has compelling reasons to endorse it. The reasons are the fruits that the deterrent effect of the principle affords, including enhanced prospects of individual survival and of the enjoyment of a safe public life. The justification of the principle does not, then, appeal to the maximization of the impersonal overall good. Rather, the idea is that the justification depends on reasons each representative law-abiding person can endorse. Some number of people will die due to police or terrorist violence, and this is permissible only if each person is less likely to die because of the policy. The relevant comparison is the counterfactual of a higher threat of death by terrorists in the absence of the policy. In the case that Hill discusses, the rationality of each person accepting the risks of the policy is manifest because deaths are compared with deaths, and policy reduces the risks for each person.

The case of killing the nonliable in warfare is somewhat less straightforward, but consider the following principle for its permissibility:

> *Permission:* A policy of responding to a just cause by a war that includes limited foreseeable but unintentional killing of people not liable to being killed is permissible only if each person has a compelling reason to accept the risks of being killed.

The idea is that a significant reduction in the risks of suffering of unjust wars is such a compelling reason. Unlike the principle for the terrorism case that Hill discusses, *Permission* does not reduce the overall risk of being killed for each person. Rather, the risk of suffering the injustices that provide just cause for war must be sufficiently reduced in order to justify the risk of being killed. The general threshold is left vague and probably must remain ever so.

[24] Hill, "Making Exceptions," 216. This formulation is mine, but I believe it is faithful to Hill's account.

But any discussions of the correct range in a particular case would seem to imply the validity of the principle.

A pacifist objection to *Permission* is that it makes people vulnerable to permissible killing who are not liable. The consideration of the reasons for and against any principle that would permit foreseeably killing people not liable to attack should appeal to generic reasons, reasons that do not gain force simply due to a particular person's location in the assignment of the benefits and burdens of the principle.[25] Assessing the pacifist challenge requires comparing the burdens of the general acceptance of *Permission* to the burdens of the general acceptance of a pacifist principle prohibiting all unintentional killing of the nonliable in war. I call this principle *Prohibition.*

Insofar as modern war inevitably involves killing the nonliable, *Prohibition* would rule out war. The threat of a belligerent response to an injustice that constitutes just cause would then no longer serve as a deterrent to the commission of the injustice if *Prohibition* were generally accepted. Foreseeably then there would be more victims of such injustice. And *Prohibition* would in some cases disallow the only means by which the just cause can be realized. It would then leave people unrescued from injustices, even if a war could rescue them. There are then two kinds of burdens of *Prohibition*, greater injustice due to less deterrence and more people trapped in injustice who could be rescued by war. Meanwhile, those who are liable to be killed, those who commit the injustices, are shielded from being killed by *Prohibition*. The principle lays a heavy burden on those who would suffer injustice and relieves a burden from those who would otherwise be permissibly targeted.

In contrast to *Prohibition*, *Permission* distributes the burden between people who may be liable to be killed, those who commit the injustice, and those who are not liable but within limits may be killed. A generic reason to prefer *Permission*, then, is that it permits adjusting the burden between those committing an injustice and those who suffer an injustice, rather than categorically protecting the former over the latter. The job of a reasonable proportionality budget is to adjust these burdens appropriately.

Both Smooth and Chunky

The justification of any proportionality budget involves a smooth feature insofar as the wrongs that would be necessary to prosecute a just cause can, in principle, cross a threshold and become excessive. This is also true

[25] Scanlon, *What We Owe to Each Other*, pp. 204–5.

according to the Personal Value Conception. However, the discussion above also involves a comparison between the categories of those who would be foreseeably killed and those who would be rescued from injustice. Any justification that distinguishes between killing and letting die is chunky. Hence, insofar as those categories matter, and not just the sheer numbers, the justification of the proportionality budget is also chunky.

Consider a formula of the following form for inclusion in the proportionality budget: S = (X)K. All the variables are positive numbers; K is the variable for the number of people unintentionally but foreseeably killed by pursuing the just cause, S is the number people saved from attack, those who would die but for the war, and X is a multiplier. So, for example, if X > 1, for every person killed more than one must be saved, so if only one person is saved for every person killed, the war is disproportional. If the limit stated in *Permission* assumes a value of X > 1, then it can be stated in general terms as follows:

> A war is proportional only if the number of people saved from injustice is greater than the number of people that the war foreseeably but unintentionally kills by a factor X, and X is greater than 1.

If X > 1, it seems implausible to complain that the formula values the lives of those killed less than the lives of those who are saved. For it would require saving more than one person to justify killing a single person.

However, perhaps one could reasonably complain that those who would not be saved because of the requirement that X > 1 are not valued equally? The idea behind the complaint is that saving a person would have been permissible if the limit included the formula that X = 1. The response to the complaint is that the failure to save someone due to the formula X > 1, someone who would be saved if the formula were X = 1, is not due to their life being valued less. Every instance of S and K is valued equally. No person's life is valued as a fraction of the value of someone else's. Rather, a person who might have been saved if X = 1, but who is not saved, is not saved because of the moral salience of killing in comparison to saving, represented by X > 1. The idea is that this difference in salience is a consequence of respect for persons. So, the difference expressed in the formula is a difference in the kind of the action, and not a difference in the value of persons. A limit on permissible killing that includes the formula with a X > 1 is consistent with valuing lives equally, while evaluating actions differently. A third possibility, in which X < 1, is totally implausible because it inverts the appropriate attitude about the moral importance of killing versus not saving.

Concluding Comments

Two conclusions seem to follow about the restrictions of wide proportionality according to the Personal Value Conception of proportionality. First, regarding the *jus ex bello* matter of continuing a war that has reached the limit of its proportionality budget, the conception does not support continuation. And second, the conception entails a proportionality budget that distinguishes between killing and failing to save, and in doing so it sets a smaller initial budget for the *jus ad bellum* than would be required if only impersonal value mattered.

In the context of justifying his opposition to the long Vietnam War, Martin Luther King, Jr. claimed: "Here is the true meaning of and value of compassion and nonviolence when it helps us to see the enemy's point of view, to hear his questions to know his assessment of ourselves."[26] I have not defended King's nonviolence. Indeed, I have suggested a reason to oppose pacifism, nonetheless the idea of taking seriously another's point of view is central to the Personal Value Conception of proportionality that I have been explicating. Moreover, if I am right, employing such a conception is particularly constraining of the harms that may be imposed in pursuit of a just cause. King finds salutary a transformation of values away from thing-orientation and toward person-orientation.[27] The arguments here agree. If proportionality is concerned with personal rather than impersonal value, it is, I believe, considerably more constraining of war.

[26] King Jr., *A Testament of Hope*, p. 237.
[27] Ibid., p. 240.

CHAPTER 9

Toward a Post Bellum *Lieber Code*

*Dan Maurer**

"Deeds done in ending" a war, according to John Rawls, reflect the aims a "liberal" or "well-ordered people" had for fighting it. Though he does not explicitly describe *jus post bellum*, Rawls does suggest that postwar peace is connected closely to the way in which war is fought and brought to a close. How states – specifically, their militaries – act will "foreshadow" the degree to which relations among the actors will be peaceful.[1] Armed forces employed to bring a measure of peace out of war often make that peace more difficult to achieve by their employment of force. To mitigate the consequences of this paradox, armed forces rely on written rules crafted by military lawyers (ultimately approved and promulgated by commanders as orders in their name) to control their use of force in two regards: to ensure troops do not violate tenets or explicit prohibitions yet in ways consistent with, and which advance, the chain of command's mission objectives.

In a *post bellum* setting, however, those rules may not be particularly useful in either of those two regards, insofar as those rules do not necessarily reflect values particularly relevant in a *post bellum* operating environment. For instance, under International Humanitarian Law (IHL), also known as the Law of Armed Conflict (LOAC), a military commander has far greater latitude to order the detention and questioning of a civilian under LOAC rules than under International Human Rights Law (IHRL) rules. Therefore, a government's broad or narrow recognition of legal constraints imposed on its military agents on foreign territory during periods of conflict is entirely a function of what legal regime it chooses to judge itself against. The LOAC framework's presumptive deference to the commander's judgment, where force need not be the last resort, and where due

* The opinions in this chapter do not reflect the official position or analysis of the Department of the Army, the United States Military Academy, or the Army Judge Advocate General Corps.

[1] John Rawls, *The Law of Peoples: With the Idea of Public Reason Revisited* (Cambridge, MA: Harvard University Press, 1999), p. 96.

process can be seen as unduly risky, adds a degree of freedom for a commander when faced with tactical problems in the field. But this level of discretion may be inappropriate in the *post bellum* environment. As the name implies, this period after conflict should be characterized by a decrease in violence, if not its complete absence, because the political objectives have changed. Those new objectives – securing peace – do not justify default deference to military commanders' choices to employ force.

This chapter attempts to lay the groundwork for a framework for *jus post bellum* conduct. I will suggest that the circumstances of soldiers on the ground, armed, and transitioning from combat to something-not-exactly combat, unintentionally risk doing what Daniel Philpott calls "spawning emotions of hatred, vengeance, and fear [leading] to further injustices and hinder[ing] the creation of just political orders."[2] These soldiers face concrete and well-worn problems in identifying *what* (if anything) constitutes a hostile act,[3] *who* (if anyone) represents a hostile actor, and *how* (if at all) to conduct themselves in the face of it. From being pelted by rocks to being sniped at from places of worship, and from facing down an oncoming motorcycle approaching a military-manned roadblock to planning missions that blend military necessities with police responsibilities, soldiers of any state face ethical dilemmas that feed and affect the *jus* of the *post bellum* period.[4] As one legal theorist has noted, however, the "traditional rules of *jus in bello* are...only partially equipped to address the problems arising in the context of peacemaking and the transition from armed conflict to peace."[5] This chapter argues that the ways in which troops are conventionally regulated in their use of force, i.e., through *jus in bello*-preserving rules of engagement, are inadequate for the preservation of *jus post bellum.* It further argues that this deficit can be overcome by commanders and the uniformed lawyers on their staffs in ways similar to those that inspired Francis Lieber's original U.S. Civil War-era "Code." Finally, it argues that overcoming this deficit, and inching toward a *post bellum* "Lieber Code," is possible by drawing principles from two distinct legal systems.

[2] Daniel Philpott, *Just and Unjust Peace: An Ethic of Political Reconciliation* (New York: Oxford University Press, 2015), p. 4.

[3] For the U.S. military's definition of "hostile act," see Chairman of the Joint Chiefs of Staff, "Standing Rules of Engagement/Standing Rules for the Use of Force," Instruction 3121.01B (June 13, 2005), Enclosure A, para. 3.d.

[4] James M. Dubik, *Just War Reconsidered: Strategy, Ethics, and Theory* (Lexington: University Press of Kentucky, 2016), p. 25.

[5] Carsten Stahn, "'*Jus ad Bellum*', '*jus in bello*' ... '*jus post bellum*'? – Rethinking the Conception of the Law of Armed Force," *European Journal of International Law*, 17, no. 5 (2006), 927.

In making my case, I will offer seven values – *justification*, *due care*, *forethought*, *nonarbitrariness*, *self-restraint*, *moderation*, and *accountability* – to act as guides in developing a tactical *jus post bellum* code. These values are small-scale repetitions of generally regarded values in two areas of legal practice: principles governing criminal law prosecutorial discretion and, second, principles of LOAC as they are generally articulated by military lawyers. Both these legal systems, in their own ways, constrain the use of force by actors invested with otherwise extraordinary authorities and licenses from the government. These systems reveal basic values in military law as they are put into practice every day by prosecutors weighing tactical decisions for when, how, and why to place a servicemember in the crosshairs of the criminal justice system, and by individual soldiers and commanders in the field weighing tactical decisions for when, how, and why to place others in more literal crosshairs. Moreover, these patterns reveal themselves in probabilities, not certainties – their very nature as indefinite norms render them relevant as standards to apply case-by-case.

These seven values can be recast as "principles" that meet a certain underacknowledged need: to design and frame conduct rules that advance the project of a *jus post bellum* environment in which armed force faces an ever-present risk of recidivism. Without such conduct rules, the very armed forces present in that *post bellum* environment with the intention of stabilizing and securing a more lasting peace ironically drive up the recidivism risk.

Jus Post Bellum and the Focus on Strategy

Jus Post Bellum is not a field lacking thoughtful normative prescriptions. However, these prescriptions explicitly emphasize *states'* goals and actions. They connect just reasons *for* war to justice resulting *from* war, arguing that, while having a *jus ad bellum* claim does not guarantee *jus post bellum*, it probably imposes an obligation to create it.[6] For example, Michael Schuck's argument in the wake of Desert Storm offered three *post bellum* principles: (1) Repentance (humility in victory, remorse for prices paid and extracted); (2) honorable surrender (in terms and in methods); and (3) restoration. These principles had two goals: to "expand the moral sensibilities of people who believe that war – while evil – is sometimes necessary for the

6 See, for example, Gary J. Bass, "Jus Post Bellum," *Philosophy & Public Affairs*, 32, no. 4 (2004), 386; and Robert E. Williams Jr. and Dan Caldwell, "*Jus Post Bellum*: Just War Theory and the Principles of Just Peace," *International Studies Perspectives*, 7, no. 4 (2006), 315.

protection of human life," and to serve as a "litmus test for the sincerity of the just war claims made before and during the conflict."[7] But it is unclear whether or how his three principles apply at levels of analysis *below* that of the state itself – do they impose obligations on political actors and strategic military planners, and on individual commanders and their troops operating in that space?[8]

Brian Orend suggests six principles for *jus post bellum*: rights vindication – that is, prescribing the goals of a postwar settlement ("to secure the basic rights whose violation triggered the justified war"); proportionality and publicity (settlement terms should be "measured and reasonable" and "publicly proclaimed"); discrimination (settlement needs to distinguish and treat differently civilians and military in the defeated state); punishment (should be proportionate, and after "fair and public" trials for war crimes); compensation (to provide financial restitution demands on the defeated may be necessary but should be discriminatory and proportionate); and rehabilitation (may include demilitarization but with the objective of rebuilding to eventually allow the defeated state to provide for its own security again).[9] In his later book, *War and International Justice: a Kantian Perspective*, however, he offers five *jus post bellum* "criteria":[10] just cause for termination, right intention, public declaration, legitimate authority and domestic rights protections, discrimination, and proportionality. But again, these principles are presented and described at the scale of state action and intent.

Michael Walzer notes that "war is a time of killing and being killed" so "the crucial requirement of *jus post bellum* is the preservation of life."[11] To that end, in the context of post war *occupations*, he offered a few examples of state-driven activities for generating *jus post bellum*: extraction of reparations; provision of basic necessities, including law and order, food and shelter; political reconstruction; special attention to protecting vulnerable populations, especially of minority groups, and ensuring the new regime is committed to that project; and retributive justice for war criminals.

7 Michael J. Schuck, "When the Shooting Stops: Missing Elements in the Just War Theory," *Christian Century*, 111, no. 30 (1994), 982–84.

8 Dubik, *Just War Reconsidered*, pp. 13, 22.

9 Brian Orend, "*Jus Post Bellum*: Perspectives of a Just War Theorist," *Leiden Journal of International Law*, 20, no. 3 (2007), 580–81.

10 Brian Orend, *War and International Justice: A Kantian Perspective* (Waterloo, Ontario: Wilfrid Laurier University Press, 2000), pp. 269–70.

11 Michael Walzer, "The Aftermath of War: Reflections on *Jus Post Bellum*," in Eric Patterson (ed.), *Ethics Beyond War's End* (Washington, DC: Georgetown University Press, 2012), 45.

At least one other military lawyer (judge advocate) has suggested his own three principle commands of *jus post bellum*, synthesizing lessons from Walzer, Schuck, and Orend:[12] seek a lasting peace; hold morally culpable individuals accountable; and extract reparations.

These principles – from Orend to Walzer – are also prescriptions for *action* – again at the political, state level – specifically for what *settlement* terms and conditions ought to occur. Orend offers his principles as guides for a "just exit strategy"[13] as if *jus ad bellum* is a prescription for a just "entry" strategy. Their arguments frame *post bellum* as a compliment to *ad bellum* – analogous bookend periods where "responsibility for fulfilling" the ethical and moral ends with ethical and moral means is "primarily political as opposed to military." They are not, however, *values* from which it would be possible to craft rules to constrain or enable appropriate military conduct under varying post-conflict conditions.[14] Soldiers on the ground are focused on more practical problems: they are repairing infrastructure, returning prisoners of war, patrolling like law enforcement, protecting government facilities and cultural sites, preventing retaliations, disarming hostile parties, training host nation security forces, and enforcing ceasefires. Applying *jus post bellum* principles as guides for state strategy is necessary, but not sufficient from their point of view. Doing so does not automatically generate tactical rules or inspire wise command decisions that enable the soldiers on the ground to exercise restraint and judgment about the use of force in ways consistent with those larger principles.

Post Bellum Paradoxes

Any *post bellum* setting is a period of ambiguities that both shape and are influenced by ongoing military operations. One brief example may serve to illustrate. In June 1999, United Nations Security Council Resolution 1244 marked the formal end of the Kosovo War by establishing a NATO Stabilization and Peacekeeping force (known as KFOR). It was decided that the responsibilities of the deployed international security presence

[12] Richard Dimeglio, "The Evolution of the Just War Tradition: Defining Jus Post Bellum," *Military Law Review*, 186 (Winter edition, 2005), 146–62; for another legal theory approach (contra moral philosophy and just war theory approach) to *jus post bellum*, see Stahn, "'*Jus ad bellum*,'" 938–41.

[13] Orend, "*Jus Post Bellum*," 581.

[14] See also Doug McCready, "Ending the War Right: Jus Post Bellum and the Just War Tradition," *Journal of Military Ethics*, 8, no. 1 (2009), 74; Louis V. Iasiello, "*Jus Post Bellum*: The Moral Responsibilities of Victors in War," *Naval War College Review*, LVII (2004), 40–51; and Williams Jr. and Caldwell, "*Jus Post Bellum*," 317–18.

in Kosovo would include, inter alia, enforcing a ceasefire; demilitarizing the Kosovo Liberation Army; and "ensuring public safety and order."[15] In 2011, when European Union police and predominantly ethnic Albanian national officials in the capital Pristina tried to establish their presence in the Serb-dominated North, Kosovo Serbs erected physical barricades at border crossings. By 2012, these roadblocks were among the last still standing. As KFOR peacekeepers worked to dismantle these barricades on several major roads, hundreds of protesters gathered, throwing stones at the troops. Gunfire erupted and the NATO soldiers, some in armored personnel carriers, responded with tear gas and small arms.[16] In the aftermath, a NATO spokesman told the media, "KFOR will not allow the situation to escalate and will use a proportional level of force necessary to maintain a safe and secure environment."[17] The mayor of the town told Reuters that KFOR had refused to allow Serb medical personnel to help wounded Serbs, and said: "A commander told me they have the authority to use deadly force on anyone who throws a stone or uses a weapon."[18] That interpretation of a rule of engagement is typical in contexts of ongoing armed conflict, where "self-defense" of authorities must be interpreted in light of *jus in bello* principles. But here, the straightforward tactical choice, ultimately made in pursuit of the unit's mission to dismantle road barricades, actually threatened to undermine the larger strategic mission announced in the UNSCR.

Episodes like this are emblematic of certain paradoxes and contradictions in the *post bellum* environment. Whether in places like postwar Kosovo, post-genocide Rwanda, Iraq after Saddam Hussein's regime fell, or certain areas at certain periods in Afghanistan, these contradictions are universal and universally confound the planning efforts of military leaders – including how, when, and under what conditions the use of force may be appropriate.

- *Contradiction 1*: Major armed hostilities have concluded, but the risk of escalation and return to fighting endures (we can call this the *Backsliding Paradox*).

[15] United Nations Security Council, "Resolution 1244," U.N. Doc. S/RES/1244 (June 10, 1999), https://undocs.org/S/RES/1244(1999).

[16] Branislav Krstic and Fatos Bytyci, "Kosovo Serbs and NATO Troops Clash in Tense North," *Reuters*, June 12, 2012. https://www.reuters.com/article/us-kosovo-serbs-idUSBRE8500HV20120601.

[17] Ibid.

[18] Ibid.

- *Contradiction 2*: The work of repair and recovery – of infrastructure, homes, businesses, governance institutions, and families torn both figuratively and literally – can begin, but that work begins sometimes under the influence or direct involvement of those parties proximately responsible for the damage (the *Community Service Paradox*).
- *Contradiction 3*: A return to the mundane practices of civil life, but doing so on roads and bridges, in homes, shops, fields, orchards, and marketplaces trampled or destroyed as collateral damage in the movement of armored vehicles and soldiers, and the impacts of bullets, white phosphorus, napalm, grenades, land mines, mortar rounds, and high-explosive armor-piercing tank shells (the "*No Place Like Home" Paradox*).
- *Contradiction 4*: The belligerents are freed to some extent from what had been a constant risk of and legal excuse for – deadly force; but the losing party often disaggregates and is absorbed back into civil society, shedding uniforms, rank, and government-backed purposes, but maybe not shedding what lawyers call their hostile means, motive, and intent (the *Suppression of Violence Paradox*).
- *Contradiction 5*: Soldiers perform reconstruction, civil affairs, diplomatic, and police investigative and patrol-like functions, relying on skills and training intended and most recently exercised in a conventional use of armed force, regulated by the LOAC's *jus in bello* principles and rules of engagement (the *Jack-of-all-Trades, Master of None Paradox*).
- *Contradiction 6*: The more that the exchange of violence between armed belligerents looks like conventional armed conflict, "military necessity" and "proportionality" arguments are easier to make and justify, and the principle of "distinction" may be less relevant depending on the locus of fighting. As a result, the rules of engagement in traditional armed conflicts offer an environment in which a soldier's individual discretion to use force is amplified or even unconstrained. If the adversary is a "declared hostile force," for example, and distinguishable from noncombatants, generally speaking the soldier does not need to wait for a hostile act or demonstration of hostile intent before using force (in self-defense) – the adversary or enemy is effectively "shoot-on-sight" unless specifically restrained by the higher command's rules of engagement.
    - In contrast, the less the exchange of violence between belligerents looks like traditional armed conflict sanctioned by nation-states,

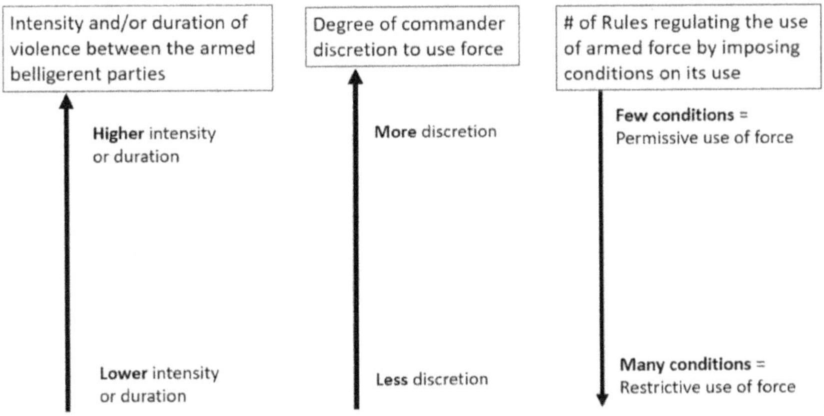

Figure 9.1 Probability and permissiveness paradox

the harder it becomes to define an attack on a target as a "military necessity," and the more significant become considerations of "proportionality" and "distinction." As a result, rules of engagement in such settings are less permissive and more restrictive of the use of force, carefully defining conditions and limitations (the *Probability and Permissiveness Paradox*). The basic idea behind the paradox is illustrated in Figure 9.1.

- *Contradiction 7*: The fact that many would plausibly claim the wars in Iraq and Afghanistan have never actually *ended*, but only changed in character,[19] underscores the chief difficulty in recognizing when we *can* or *should* consider the use of armed force to be grounded in *post bellum* ends, ways, and means. As one commentator noted, the timeframe of *jus post bellum* can be the "period following major combat operations up to and after the cessation of hostilities. This means that *jus post bellum* considerations are likely to overlap *jus in bello* [considerations]" (the *Drawing the Line Paradox*).[20]

Though the paradoxes are commonly known, these contradictions are not systematically accounted for when planning and designing practical rules of engagement. This is unfortunate, for the opportunity to do so

[19] Jay S. Bybee, "Authority of the President Under Domestic and International Law to Use Military Force Against Iraq," *Opinions of the Office of Legal Counsel*, 26 (2002), 10–13 (*Authority of the President Under Domestic and International Law to Use Military Force Against Iraq*), https://fas.org/irp/agency/doj/olc/force.pdf.

[20] McCready, "Ending the War Right," 67.

certainly exists. Lawyers in the U.S. military play an "advise and assist" role in two ways. First, they counsel commanders on the disciplinary options they possess and managing that process – this may include prosecuting soldiers accused of violating some positive law or rule, including allegations of war crimes, under the Uniform Code of Military Justice (UCMJ) and the U.S. War Crimes Statute.[21] Second, they draft, explain, persuade, and train commanders, staff, and troops on rules of engagement.[22] In the form of legally enforceable orders, such rules are designed to reduce for the soldier those situations that generate ethical dilemmas of the sharpest kind: uncertainty about when, where, why, against whom, and how to employ the means and methods of warfare with which they have been entrusted.[23] Nevertheless, to borrow some jargon from the law, military lawyers do not have anything approximating binding precedent from the *post bellum* jurisdiction from which to clearly deploy legal arguments to help commanders shape the behavior of armed forces after the cessation of hostilities but before an acceptable degree of peace makes their presence no longer necessary. Without binding precedent, military lawyers must draw from persuasive "secondary" authorities to account for these paradoxes.

Jus Post Bellum Has Not Been Part of the Discussion on the Ground

As a "phase" of conflict that may or may not deserve its own, distinct, ethical considerations and its own legal constraints, *jus post bellum* has not yet graduated fully into the military *lawyer* community's collective awareness and lexicon.[24] The term does not appear in, and the concept is largely absent from, the U.S. Department of Defense's *Law of War Manual*.[25] The closest it gets is the discussion of the application of occupation law

[21] The UCMJ is Codified at 10 U.S.C. § 801–946a. The US War Crimes Statute is 18 U.S.C. § 2441.

[22] Frederic L. Borch, *Judge Advocates in Combat: Army Lawyers in Military Operations from Vietnam to Haiti* (Washington, DC: Office of the Judge Advocate General and Center for Military History, 2001), p. 313; *United States Army Field Manual*, "Legal Support to the Operational Army," FM 1–04 (2013), https://bit.ly/3JevWfn, p. 7–2 (para. 7–8 and 7–9).

[23] U.S. Department of Defense, *Legal Support to Military Operations, GL-3*, Joint Publication (JP) 1–04 (Washington, DC: United States Department of Defense, 2011).

[24] See Eric De Brabandere, "The Responsibility for Post-Conflict Reforms: A Critical Assessment of Jus Post Bellum as a Legal Concept," *Vanderbilt Journal of Transnational Law*, 43, no. 1 (2010), 121–22. However, see Dimeglio, "The Evolution of the Just War Tradition," 116; and see Stahn, "'*Jus ad Bellum*,'" 927.

[25] U.S. Department of Defense, *Law of War Manual* (Washington, DC: Office of General Counsel, United States Department of Defense, 2016). Cited hereafter as *Law of War Manual*.

to "situations not constituting belligerent occupation." There, the U.S. government policy is that this body of law does *not* apply, "as a matter of law," unless certain conditions are met, and does not apply in "post-war situations." In other words, occupations exist within the timeline of the period of war, and occupation law (from the Hague Regulations to the Fourth Geneva Convention) applies only during International Armed Conflicts. The manual does, however, acknowledge that occupation law *might* usefully apply some of its rules by analogy in those post-conflict situations.[26] Even more surprisingly, the new Army and Marine Corps Field Manual, *The Commander's Handbook on the Law of Land Warfare*, which took two decades to draft, staff, and publish, defines both *jus ad bellum* and *jus in bello*, but does not say a word about *jus post bellum*. The closest it gets are brief references to war crimes tribunals,[27] "post-conflict battlefield cleanup" (the duty of environmental remediation of the "remnants of war"[28]), post-conflict return of human remains,[29] and "relations of forces with local populations during an armistice."[30]

It is an open question in international law as to whether the Law of Occupation – either customary or under the Fourth Geneva Convention – applies in situations after hostilities have officially and formally ended but in which armed resistance to the occupier, or any armed force present, remains.[31] The official U.S. position is that IHRL – while a default baseline of obligations a state owes to its own citizens in its own territory – is subsumed by IHL or LOAC because at a time of armed conflict, the *lex specialis* of IHL is the only valid international law controlling and legitimizing the use of state-sanctioned violence and protection of victims of that violence. That is to say, if the U.S. forces follow IHL or LOAC, they are satisfying the demands of IHRL in that specific time and place for the duration of that armed conflict.[32] But this is far from the settled or consensus view. For U.S. practitioners at least, it is not clear whether IHL is always or necessarily subsumed within IHRL, or what legal regime works better (or should work at all) in ever more common nontraditional wars

[26] Ibid., p. 759, para. 11.1.3.

[27] U.S Department of Army, *Field Manual 6–27, The Commander's Handbook on the Law of Land Warfare* (Washington, DC: Headquarters, Department of the Army, 2019), pp. 8–10, para. 8–57.

[28] Ibid., pp. 2–23, para. 2–144.

[29] Ibid., pp. 4–4, para. 4–15.

[30] Ibid., pp. 7–8, para. 7–65.

[31] See Jack L. Goldsmith III, "'Protected Person' Status in Occupied Iraq Under the Fourth Geneva Convention," *Opinions of the Office of Legal Counsel*, 28 (2004), 37n4, https://bit.ly/3QmKLzg.

[32] See *Law of War Manual*, p. 9n13, and Christopher Greenwood, "Historical Development and Legal Basis," in Dieter Fleck (ed.), *The Handbook of Humanitarian Law in Armed Conflicts* (1999), 9.

spread across multiple sovereign states, involving nonstate actors.[33] Nor is it clear or definitive when to consider hostilities sufficiently complete to no longer believe that traditional LOAC and *jus in bello* concepts apply.[34]

We are still left with serious, open normative questions: should we follow by custom and analogy those protections afforded under the Law of Occupation even in contexts where we are not occupiers?; should we follow by custom and analogy those protections afforded under Common Article 3 to situations after relatively-larger scale belligerency has ended but troops remain?[35] Questions about *who* would constitute a "lawful combatant" or an "unlawful, unprivileged, belligerent" or even a simple "hostile actor," are likely to be determined conflict- by-conflict, party-by-party.[36] Conditions in that postwar, pre-peace operating environment complicate who should be designated as beneficiaries of "collective self-defense,"[37] and they complicate troops' resort to using force in self-defense because the meaning of a hostile actor's "imminent use of force"[38] is probably a function of that actor's reduced fighting capacity and capability, coupled with uncertainty about their intent. Thus "direct participation in hostilities" is an even greater puzzle to define in this post-conflict, pre-peace time and

[33] See, for example, Dietrich Schindler, "Human Rights and Humanitarian Law: Interrelationship of the Laws," *American University Law Review*, 31 (1982), 935; Ariel Zemach, "The Unpleasant Responsibilities of International Human Rights Law," *Denver Journal of International Law & Policy*, 38 (2010), 421; Alexander Orakhelashvili, "The Interaction Between Human Rights and Humanitarian Law: Fragmentation, Conflict, Parallelism, or Convergence?," *European Journal of International Law*, 19, no. 1 (2008), 161.

[34] See, for example, Laurie R. Blank and Benjamin R. Farley, "Identifying the Start of Conflict: Conflict Recognition, Operational Realities and Accountability in the Post-9/11 World," *Michigan Journal of International Law*, 36, no. 3 (2015), 496–518.

[35] The United Nations peacekeepers are obligated to follow "fundamental principles and rules of international humanitarian law" whenever they are "actively engaged" in armed conflict in which they have a right to exercise self-defense as lawful combatants. *Secretary-General's Bulletin*, ST/SGB/1999/13 (New York: United Nations, 1999). The substantial majority of those promulgated rules are indistinct from generally prescribed rules applicable in most forms of conflict and reflected in various published rules of engagement. See, for example, *Desert Storm Rules of Engagement Card (Iraq, 1991)*, in Chairman of the Joint Chiefs of Staff Instruction, *Standing Rules of Engagement* (however, many of the ROE prohibitions articulated for that kinetic international armed conflict carried the caveat of "unless necessary to save US lives," which has no parallel in the United Nations bulletin). See also Hamdan v. Rumsfeld, 548 U.S. 557 (2006), www.supremecourt.gov/opinions/05pdf/05-184.pdf (slip opinion, at 66–70) (holding that Common Article 3 of the Geneva Conventions governs the United States conduct in Afghanistan during the so-called War on Terror).

[36] Ex parte Quirin, 317 U.S. 1, 31 (1942). For how the US defines these classes and categories of actors, see *Law of War Manual*, p. 102–3 (para. 4.3).

[37] For the United States military's definition of "collective self-defense," see Chairman of the Joint Chiefs of Staff Instruction, *Standing Rules of Engagement*, Enclosure A, para. 3.c.

[38] On the U.S. military's meaning of "imminent," see ibid., para. 3.g.

place.[39] Ultimately, the difficult question is: to what degree and under what conditions may force be used by those armed forces when the political and military objective is to return to *status quo ante*, or conversely create a more durable, secure peace?[40]

Right Conduct After War but Before Peace?

As a lawyer advising commanders about what rules they ought to impose to manage the use of force to accomplish their missions, I am beset by the seven paradoxes described above. The challenge is compounded because the moral and ethical principles for *jus post bellum* that might focus my recommendations are, at best, murky. Fortunately, this situation is not unprecedented.

In 1952, in a particularly important opinion about a president's constitutional discretion, United States Supreme Court Justice Robert Jackson reflected on his time, before the Court, working for the executive branch. He wrote that there is a "poverty of really useful and unambiguous authority applicable to concrete problems of executive power as they actually present themselves."[41] Jackson served earlier as the lead U.S. prosecutor for the war crimes tribunals at Nuremberg, taking a major part in what is often the most visible and contentious area of *jus post bellum* activity.[42]

Justice Jackson's observation is similar in spirit to Walzer's comment, in his preface to *Just and Unjust Wars*, that when philosophers seek an "engagement" with the "most profound questions of moral philosophy" they "often miss the immediacies of political and moral controversy and provide little help to men and women faced with hard choices."[43] Jackson's comment – made with respect to the seemingly boundless and constitutionally supported prerogative of an American president during the Korean War – is also noteworthy because it served as a premise for his attempt to craft just such an analytical guide. He hoped it would be "applicable to concrete problems of executive power as they actually

39 On the U.S. military's interpretation of "Direct Participation in Hostilities" [DPH], *see Law of War Manual*, § 5.8.3. The United States officially supports the principle behind, but not the text of, the Geneva Conventions Additional Protocol I, Article 51(3) definition. *Protocol Additional to the Geneva Conventions of 12 August 1949, and relating to the Protection of Victims of International Armed Conflicts* (Protocol I), 8 June 1977.

40 Walzer, "The Aftermath of War," 37.

41 Youngstown Sheet & Tube Co. v. Sawyer, 343 U.S. 579, 634 (1952) (Jackson, J., concurring).

42 Orend, "*Jus Post Bellum*," 574.

43 Michael Walzer, *Just and Unjust Wars: A Moral Argument with Historical Illustrations* (New York: Basic Books, 2015), p. xxix.

present themselves." Constitutional text and philosophical arguments from Alexander Hamilton in *Federalist no. 70* are one thing;[44] but the gritty politics of compromise, confusion, collaboration, and requirements for less-than-deliberative decision-making are another. He presented a case to manage what one might call the "fog and friction of politics." His goal was a framework that might better signal, to all three branches of government, the more nuanced shape, pockmarked with caveats, of that presidential discretion under those less-than-ideal circumstances.

Jackson's objective is comparable to the objective spurring some of the procedures and rationales we use to employ or legitimize military actions before, during, and after the military is employed in conflict – a context also thick with compromise, confusion, and high stakes, one requiring discretion and judgment, under restrictive time constraints, public visibility, and without complete information. Today, these procedures and rationales are written, published, distributed, and trained as *rules of engagement, rules for the use of force*, or *tactical directives*. They are formally from the commanders (though, in the U.S., judge advocates usually play significant roles in drafting them), and based on considerations of their mission, their operating environment, the threat, the danger to their own troops, and the larger strategic, public, and policy considerations. They owe their existence, as lineal descendants, to Francis Lieber and his code, General Order 100.[45] As John Fabian Witt wrote, this project – initiated during the U.S. Civil War by a law professor at the request of President Abraham Lincoln, his Secretary of War, and the Union's General-in-Chief (himself a lawyer) – was "breathtakingly ambitious." Lieber's Articles articulated rules for right conduct by soldiers and commanders in the field, but also the "rationales and general principles that lie behind the rules. It is a working document for the soldier and the layman, not a treatise for the lawyer or the statesman."[46] The code's purpose was more than a political commitment at the nation-state level to follow customary international law of war in what was becoming, to everyone's surprise, one of the bloodiest civil wars in history. It was also direct guidance to commanders and soldiers on thorny and then-novel subjects like armed and unarmed civilian Confederate sympathizers and escaped slaves on or near the battlefield.

[44] Alexander Hamilton, "Federalist No. 70," in Clinton Rossiter (ed.), *The Federalist Papers* (New York: Penguin, 2003), pp. 421–428.

[45] *Instructions for the Government of Armies of the United States in the Field* (1863), reprinted in John Fabian Witt, *Lincoln's Code: The Laws of War in American History* (New York: Free Press, 2012), Appendix.

[46] Witt, *Lincoln's Code*, p. 2.

The Lieber Code's animating goal and many of its key ideas influenced the Hague Regulations in 1899 and again in 1907.[47] Of particular note is Lieber's Article 16, on the limits of "military necessity," where he writes that it "does not include any act of hostility which makes the return to peace unnecessarily difficult."[48] Thus, even this early incantation of rules of engagement, derived from customary international law, envisions a close causal relationship between what happens *during* conflict and what results *from* the conflict: in other words, the means and methods of warfare may paradoxically disable and frustrate the ends and goals of the war.

Of course, none of these considerations – mission, threat, risk, and policy – exist as independent or static. They are interdependent and subject to change. This makes the explicit definitions and scoping of conduct by military lawyers and planners operating in combat or combat-like conditions an exercise not unlike a child blindly swinging at a piñata with a stick, encircled by their young friends and slightly concerned parents. In other words, there is a significant dose of hope that the means and methods are close enough to the mark that military forces do not inadvertently cause additional and unnecessary chaos, danger, uncertainty, and fear precisely when such chaos should settle, such danger should diminish, such uncertainty should give way to positive expectations, and such fear should be outweighed by a sense of relative safety. We can simply recast Jackson's phrase to apply it by analogy here: "[t]hat comprehensive and undefined [military] powers hold both practical advantages and grave dangers for the country will impress anyone who has served as legal adviser to a [commander] in time of transition and public anxiety."

These typical *post bellum* challenges and paradoxes reveal why achieving *jus post bellum*, when armed forces remain present after the conflict (and actively employed in peace-building, peacekeeping, and peace-enforcement), needs special considerations that have been so far neglected by both the literature and the practical application of the IHL "in the field." These considerations should, as Jackson's did for his purposes in *Youngstown*, account for the public visibility, compromise, confusion, collaboration, discretion, and judgment under restrictive time constraints, without complete information, that are the hallmarks of any *post bellum* operating environment.

[47] Jordan J. Paust, "Dr. Francis Lieber and the Lieber Code," *Proceedings of the Annual Meeting (American Society of International Law)*, 95 (2001), 112–15; see also Prosecutor v. Tadić, Decision on Defence Motion on Jurisdiction, para. 63 (Int'l Crim. Trib. for the Former Yugoslavia Aug. 10, 1995).

[48] Witt, *Lincoln's Code*, p. 377, Appendix, reprinting U.S. War Department, Adjutant General's Office; General Orders, No. 100 (1863).

Post Bellum Premises

This chapter has described several paradoxes of the *post bellum* operating environment and has described Lieber and Jackson in their respective fields working to establish principles for guiding action in the real world. But it is sensible to establish common agreed-upon premises before we can fully justify the search for the applicable values. I identify five such premises. First, experience reveals, and theorists recognize, no bright line demarcating a "phase transition" from an operating environment where *jus in bello* concepts (including noncombatant immunity) apply to one in which they no longer apply.[49] Second, the armed forces of a belligerent party often remain armed and available to apply force in localities after hostilities have officially or practically terminated and where residual conflict remains reasonably foreseeable, but where use of force is not their primary mission or activity. Third, military lawyers act as "agents" for their commanders (as "principals") in two senses: first, as "military justice strategists"[50] in their assigned function as personal, special, confidential staff advisers; second, as LOAC experts on the commander's planning staff,[51] working alongside tacticians, logisticians, medical planners, and intelligence analysts. Fourth, in a *post bellum* operating environment, where residual armed conflict is reasonably foreseeable and a return to peace is uncertain, the lawyer-as-agent (in both senses) is likely to seek a sensible, manageable framework for considering how, where, when, why, and with what means military force could be applied that reduces the commander's risk of breaching legal standards, prohibitions, and accepted norms *without* imposing additional undeserved tactical and operational risk to their forces or the mission. As Gary Bass put it, there is a need "to temper the righteous and discourage the reckless."[52] Finally, just like Robert Jackson's context when he wrote his opinion in 1952 about presidential authority, there is no commonly accepted *jus post bellum* set of principles, standards, or rules (either prescriptive, proscriptive, or aspirational) for commanders and individual servicemembers at the tactical and operational level.

49 McCready, "Ending the War Right," 75; and see Orend, *"Just Post Bellum,"* 573, and Stahn, "'*Jus ad Bellum*,'" 923 (calling it the "erosion of the classical war-peace distinction").

50 Dan Maurer, "The Judge Advocate as Strategist," *Small Wars Journal*, January 28, 2017, https://smallwarsjournal.com/jrnl/art/the-judge-advocate-as-strategist.

51 See, for example, U.S. Department of Defense, *Legal Support to Military Operations* (especially chaps II and III).

52 Bass, "Jus Post Bellum," 385. Or to paraphrase Daniel Philpott, aiming for the right conduct and reacting rightly "in response to wrong conduct." See also Philpott, *Just and Unjust Peace*, p. 53.

There are undesirable consequences of accepting and acknowledging these premises: Armed forces may reasonably (according to their subjective point of view) but inappropriately (according to our later, objective, point of view) rely on (a) *jus in bello* principles and the published rules of engagement (ROE), or (b) some *ad hoc* collection of rules or informal nonprescriptive methods and procedures deemed necessary at the time under those circumstances.[53] As a foreseeable result, the armed forces in post-conflict settings create unnecessary risk or stressors on *jus post bellum* reconciliation, recovery, and return efforts.

If, as mentioned earlier, nothing in military doctrine, law, or policy definitively situates LOAC in a *jus post bellum* environment, then from what source might we derive such a set of principles, standards, and rules for regulating, checking, training, criticizing, and interpreting the *use* of armed force in a setting that is somewhere between war and peace? In other words, if we wished to create a *jus post bellum* Lieber Code, where do we start? It turns out that military commanders, their staff planners, and their legal advisers need not face the choice between framing their own military conduct in a *post bellum* setting under the more permissive, commander-discretion and military necessity-driven LOAC and the more restrictive observance of IHRL.

Seven Themes of Prosecuting Crime and War – the Common Values

There is clear reason to think critically about armed forces and their conduct *post bellum*, but there is no clear reason why we should be thinking of the *post bellum* environment *only* in terms of state-level actions and prewar, *jus ad bellum* claims. Earlier, we likened these other efforts to articulate guiding principles for *jus post bellum* as strategies for a state's *post bellum* actions, it follows that we should discuss what *jus post bellum* demands at a tactical level.

Given that military lawyers must contend with this challenge, and given that they populate this environment by both prosecuting soldiers accused of misconduct and advising commanders through the thicket of international law during the planning and execution of military operations, it seems sensible to accept Orend's implied invitation to see parallels in

[53] Informal, nonprescriptive methods are usually referred to, in the military, as "tactics, techniques, and procedures" (usually shortened to "TTPs"). See United States Department of Defense, *Department of Defense Dictionary of Military and Associated Terms (as amended)*, Joint Publication (JP) 1–02 (Washington, DC: Department of Defense, 2016), A-183 (for acronym), 235 (definition of "tactics"), 239 (definition of "techniques"), and 190 (definition of "procedures").

criminal law and law of war. He writes: "there is no inconsistency, or mystery, in holding particular actors, in complex local conflicts, to more general, even universal, standards of conduct. Judges and juries do that daily, evaluating factual complexities of a given case in the light of general moral and legal principles."[54]

When opening the practice of these two fields to such scrutiny, the parallels between criminal prosecution on the one hand, and the *jus in bello* principles of distinction, military necessity, proportionality, and unnecessary suffering (and perhaps even a principle of precaution) on the other, become readily apparent. Under the American form of military justice, criminal prosecutorial-like discretion is in the hands of commanding officers, aided and advised by licensed attorneys (judge advocates). The commander, not the lawyer, has the legal duty to maintain good order and discipline among their troops in order to keep the unit ready to perform its mission. It is traditionally assumed that nobody is better positioned than the commander to know the effect of a soldier's wrongdoing, or potential punishment, will have on that commander's mission.[55] Therefore, not only are commanders allocated "tactical" authority to conduct law enforcement-like searches and seizures,[56] initiate investigations,[57] order pretrial confinement,[58] and execute the military equivalent of arresting and indicting,[59] the law also allows them to choose whether to pursue a court-martial or opt to address misconduct in other less-stigmatizing administrative ways.[60] Like standard principal-agent relationships, the commanders rely on the "expertise, responsibility, and corporateness"[61] of the judge advocate legal counsel to advise on appropriate dispositions, and, if charging a servicemember with a crime, to represent the U.S. government in the court-martial as prosecutor.

54 Orend, "*Jus Post Bellum*," 577.

55 Orloff v. Willoughby, 345 U.S. 83, 93–94 (1953); see generally Parker v. Levy, 417 U.S. 733 (1974); see Walter T. Cox et al., "Report of the Commission on the 50th Anniversary of the Uniform Code of Military Justice," 4 November 2001, https://bit.ly/3SJzZEJ (hereafter cited as 2001 Cox Commission Report); 10 U.S.C. § 3583, Pub. L. 105–85, §507(a)(1), 111 Stat. 1726 (1997).

56 U.S. Department of Defense, *United States Manual for Courts-Martial* (Washington, DC: United States Department of Defense, 2019), Military Rules of Evidence 315 and 316, pp. III-14–III-16. Hereafter cited as *Manual for Courts- Martial.*

57 Ibid., Rule for Courts-Martial 303, p. II-21.

58 Ibid., Rule for Courts-Martial 305, pp. II-23–II-27.

59 Ibid., Rules for Courts-Martial 302, p. II-19; 307, p. II-29, and 407, p. II-48.

60 Ibid., Rules for Courts-Martial 306, p. II-27; *see* 10 U.S.C. § 877–934 (Articles 77–134, UCMJ).

61 Samuel P. Huntington, *The Soldier and the State: The Theory and Politics of Civil-Military Relations* (Cambridge, MA: Harvard University Press, 2003 [1957]), p. 8.

The sources for these patterns in criminal law are many: the *Manual for Courts-Martial*; regulations on judge advocates' professional responsibility and conduct;[62] codes of conduct and professional ethics published by the jurisdictions in which judge advocates are licensed to practice law;[63] the American Bar Association's and the Department of Justice guidance and standards for prosecutors.[64] Moreover, just like learning lessons from previous conflicts and trying to adapt or apply those to new situations, lawyers can turn to case law in which appellate courts have reviewed and commented on the actions of prosecutors at trial and commanders before and after trial.[65]

The sources for these patterns in LOAC include customary international law, treaty obligations, and (for U.S. military lawyers) the Defense Department's voluminous *Law of War Manual*, along with the unclassified *Standing Rules of Engagement/Standing Rules for the Use of Force*.

Seven patterns appear to emerge from these various rules shared by both fields – prosecuting military crime, and prosecuting hostilities.

(1) *Justification* – having a legitimate, valid (i.e., defensible) reason for making a choice *to* act or not.

In a criminal prosecution, Justification is manifested by a commitment to only prosecuting an offense when its nature, seriousness, circumstances, and the actor's culpability warrant it,[66] when the government has *sufficient evidence* to reasonably believe a crime has been committed,[67] a reasonable belief that prosecution *will result* in conviction by a beyond a reasonable

62 U.S. Department of Army, *Rules of Professional Conduct for Lawyers* (Washington, DC: United States Department of the Army, 2018), Regulation 27–26.

63 See, for example, Supreme Court of Ohio, *Ohio Rules of Professional Conduct* (Columbus, OH: Ohio Supreme Court, 2020), https://bit.ly/3bTYdLN.

64 American Bar Association [ABA], *Criminal Justice Standards for the Prosecution Function* (Washington DC: American Bar Association,2017), https://bit.ly/3QBOAQI; U.S. Department of Justice, *Justice Manual*, Title 9, § 9–27.000, "Principles of Federal Prosecution," www.justice.gov/jm/jm-9-27000-principles-federal-prosecution.

65 See, for example, United States v. Howe, 37 M.J. 1062 (N-M Ct. of Crim. App. Review 1993); United States v. Dancy, 38 M.J. 1 (C.M.A. 1993); United States v. Meek, 44 M.J. 1 (C.M.A. 1996); United States v. Barry, 78 M.J. 70 (C.A.A.F. 2018); and United States v. Boyce, 76 M.J. 242 (C.A.A.F. 2017).

66 See, for example, ABA, *Criminal Justice Standards for the Prosecution Function* (2017), 29, Standard 3–4.4 ("Discretion in Filing, Declining, Maintaining, and Dismissing Criminal Charges"); and *Justice Manual*, at § 9–27.230.

67 ABA, *Criminal Justice Standards*, 29, Standard 3–4.3(a) ("Minimum Requirements for Filing and Maintaining Criminal Charges"); U.S. Department of Army, *Rules of Professional Conduct for Lawyers*, 68 (Rule 3.8(a)), and *Manual for Courts-Martial*, appendix 2.1 (nonbinding disposition guidance), pp. A2.1–2, para. 2.3.

doubt standard,[68] and the *decision to charge* is in the interests of justice.[69] Justification is further manifested (for military justice) as due consideration of the effect of the offense on good order and discipline of the command.[70] In LOAC, justification encapsulates the *jus in bello* criterion of military necessity (use that is necessary to defeat the enemy as quickly and efficiently as possible, as long those means and methods do not violate the law of war),[71] which itself is considered to be a case-specific, circumstance-specific conditional application.[72]

> (2) *Due care* – the exercise of attention, precision, accuracy, and caution "in the attack" be it in charging a fortification or charging a soldier with a crime.

In criminal prosecution, due care means fully investigating the alleged offense with diligence and thoroughness,[73] considering the strength and admissibility of the evidence,[74] and charging the right suspect in the narrowest way possible consistent with the suspect's actions and intentions.[75] It means ensuring the defense has access to any evidence that could cancel or reduce the accused's culpability.[76] In LOAC, due care means acting with sufficient precautionary measures in place to avoid unnecessary suffering and unnecessary collateral damage,[77] and distinguishing between combatant and noncombatant, and between protected and unprotected objects.[78]

[68] ABA, *Standards for the Prosecution Function*, Standard 3–4.3(a), and *Justice Manual*, § 9–27.300 (Comment).

[69] *Justice Manual*, § 9–27.220 and § 9–27.230 (comment). While this may be the federal standard, many states permit charging based solely on probable cause. Conviction, however, is always premised on proof beyond a reasonable doubt.

[70] *Manual for Courts-Martial*, appendix 2.1 (nonbinding disposition guidance), pp. A2.1–2, para. 2.1.a.–n.

[71] North Atlantic Treaty Organization [NATO], *Glossary of Terms and Definitions*, AAP-6 (Brussels: NATO Standardization Office, 2019), 82.

[72] *Law of War Manual*, p. 56 (para. 2.2.3).

[73] ABA, *Standards for the Prosecution Function*, 9, Standard 3–1.9 ("Diligence, Promptness and Punctuality"); compare Supreme Court of Ohio, *Ohio Rules of Professional Conduct* (2020), Rule 1.3.

[74] See, for example, *Manual for Courts-Martial*, appendix 2.1 (nonbinding disposition guidance), pp. A2.1–2, para. 2.1.h.

[75] Ibid., para. 2.4.

[76] Brady v. Maryland, 373 U.S. 83 (1963); compare Supreme Court of Pennsylvania, *Rules of Professional Conduct*, Rule 3.8(d) (Harrisburg, PA: Supreme Court of Pennsylvania, 2019), https://bit.ly/3QkUi9L; with West Virginia Judiciary, *Rules of Professional Conduct*, Rule 3.8(d) (Charleston, SC: West Virginia Judiciary, 2015), https://bit.ly/3w1zw72.

[77] See, for example, *Law of War Manual*, p. 190 (para. 5.2.3); see also Michael J. Matheson, "Remarks on the United States Position on the Relation of Customary International Law to the 1977 Protocols Additional to the 1949 Geneva Conventions at the Sixth Annual American Red Cross-Washington College of Law Conference on International Humanitarian Law (Jan. 22, 1987)," *American University Journal of International Law and Policy*, 2, no. 2 (1987), 426–27.

[78] *Law of War Manual*, p. 62 (para. 2.5).

(3) *Forethought* – the use of military force or military criminal jurisdiction as an instrumentality of command is considered and reviewed for prudence and lawfulness *before* taking action.

In criminal prosecution, Forethought means considering the availability and willingness of key witnesses, or victims, to testify;[79] the accused's willingness to cooperate;[80] the input from law enforcement investigators;[81] the probable sentence or other collateral consequences of a conviction;[82] and the mission-related responsibilities of the commander and the effect of the prosecution on the unit and its readiness.[83] In LOAC, it means planning so that any force used will be proportional force and no more,[84] on positively identified lawful targets only,[85] and only for objectives that are militarily necessary – to "refrain from attacks in which the expected harm incidental to such attacks would be excessive in relation to the concrete and direct military advantage anticipated to be gained and to take feasible precautions in planning and conducting attacks to reduce the risk of harm to civilians and other persons and objects protected from being made the object of attack."[86]

(4) *Nonarbitrariness* – rules and exercises of authority granted by those rules are consistent and applied nondiscriminatorily; they are not *ad hoc*, *ex post facto*, groundless, or unreasonably discriminatory or preferential.

In criminal prosecution, nonarbitrariness means not deviating from the demanding standards and explicit rules of due process,[87] to apply the law and seek punishment that is "warranted, appropriate, and fair" but constrained by a known menu of legitimately authorized choices.[88] In LOAC,

79 *Manual for Courts-Martial*, appendix 2.1 (nonbinding disposition guidance), pp. A2.1–2, para. 2.1.g.
80 Ibid., para. 2.1.k.; *see also Justice Manual*, § 9–27.230 (comment).
81 *Manual for Courts-Martial*, appendix 2.1 (nonbinding disposition guidance), pp. A2.1–2, para. 2.1.i.; see also ABA, *Standards for the Prosecution Function*, 20–21, Standard 3–3.2 ("Relationships with Law Enforcement").
82 *Manual for Courts-Martial*, appendix 2.1 (nonbinding disposition guidance), pp. A2.1–2, at para. 2.1.m.
83 Ibid., at para. 2.1.a. and c.
84 *Law of War Manual*, p. 62 (para. 3.5) (warning the practitioner to not "conflate" the principle of proportionality as applied in *jus ad bellum* with that of proportionality in *jus in bello* contexts; nevertheless, the common denominator for both is a comparison of expected harm and expected gain).
85 Gary D. Solis, *The Law of Armed Conflict: International Humanitarian Law in War* (New York: Cambridge University Press, 2016), p. 492.
86 *Law of War Manual*, p. 61 (para. 2.4.1.2).
87 For example, U.S. Constitution, Fourth, Fifth and Sixth amendments; and United States v. Katz 389 U.S. 347 (1967).
88 The phrase "warranted, appropriate, and fair" described the commander's ideal "disposition decision" using their authority under the Uniform Code of Military Justice and was described with a list of considerations in Rule for Courts-Martial 306(b) in the *Manual for Courts-Martial* (2016). When

nonarbitrariness means consistently trying to abide by the principles of *jus in bello*, encoding them by deriving specific conduct constraints called rules of engagement,[89] that have built-in discretionary flexibility under certain conditions, but are binding on those within the command's "jurisdiction."

> (5) *Self-restraint* – in deciding *if* to use force, only use force or criminal law for certain, defined and clearly articulated reasons and under certain predictable conditions.

In a military criminal prosecution, self-restraint means making a choice to prosecute only when evidence and the circumstances support it,[90] and not being influenced by the characteristics of the accused like race, gender, sexual orientation, politics, their noncriminal activities or associates, or by political pressures or unreasonably swayed by personal feelings concerning the accused, the victim, or the crime.[91] In LOAC, self-restraint is demonstrated by the use of force only to the extent that it is necessary to quickly and efficiently overcome the capabilities and willpower of the enemy;[92] it is further demonstrated by not allowing military means or methods to run rampant, causing unnecessary suffering;[93] and abiding by prohibitions on certain conduct that often surfaces from combat's inherent confusion, fear, distrust, ambiguities, danger, and time constraints.[94]

> (6) *Moderation* – temper the use of force or the law; just because an extreme application is lawful and within the ambit of your discretion, does not mean it must be followed (we often shorthand our advice to commanders as "lawful, but awful").

In criminal prosecution, moderation means adjusting one's prosecutorial strategy and tactics to remain open to less punitive outcomes, like

the most recent version of the manual was published in 2019, incorporating the most significant substantive and procedural changes to military justice in decades in accordance with the Military Justice Act of 2016 (in the National Defense Authorization Act of 2017, Pub. L. 114–328, codified as amended in scattered sections of 10 USC) the phrase inexplicably disappeared but was replaced by a much more detailed "guidance" to commanders and their lawyers in considering how to promote and enforce good order and discipline. *Manual for Courts-Martial*, appendix 2.1 (nonbinding disposition guidance), pp. A2.1–2, at para. 1.1.b. and 1.2.a.

89 US Army, *Operational Law Handbook*, 18th edn. (Charlottesville, VA: The Judge Advocate General's Corps Legal Center and School, 2018), p. 81; see, for example, Chairman of the Joint Chiefs of Staff, *Standing Rules of Engagement*.

90 See, for example, ABA, *Standards for the Prosecution Function*, 29, Standard 3–4.3.

91 *Manual for Courts-Martial*, Appendix 2.1 (nonbinding disposition guidance), pp. A2.1–2, at para. 2.7a.–e.; and *Justice Manual*, § 9–27.260 (initiating and declining charges – impermissible considerations).

92 *Law of War Manual*, p. 58 (para. 2.3).

93 Ibid., pp. 357–64 (para. 6.6).

94 Ibid., p. 8 (para. 2.3.1).

mediation,[95] plea agreements, and reduced sentences,[96] and arguing for punishments that fit not just the crime but also fit all the *circumstances* of the crime.[97] In LOAC, moderation is demonstrated by the pursuit of non-aggressive, or nonkinetic, nonlethal tactics, techniques, and procedures to achieve the desired effect, even if slower, more cumbersome, and below the threshold of legally proportional use of force.[98]

> (7) *Accountability* – laws in both settings impose enforceable obligations on commanders, reporting responsibilities, and venues for investigating, prosecuting, and punishing offenders.

In criminal prosecution, accountability is in the establishment of feasible, practical settings in which due process is permitted to unfold, notwithstanding demands of time, location, or public scrutiny: Lawyers and law enforcement are held to account for their mistakes, either by way of evidence being excluded;[99] a prosecution prevented from going forward;[100] a duty upon receiving clear and convincing evidence that an accused was convicted of a crime they did not commit to "remedy the conviction" by disclosing the evidence to the appropriate court;[101] or investigation and discipline meted out by their respective licensing bar authority.[102] In LOAC, accountability means ROE and law of war violations are reported through the chain of command or external oversight or observer,[103] to

95 For example, the City of Columbus, OH, Mediation Program (www.columbuscityattorney.org/prosecution-mediation.aspx); Idaho Supreme Court, "Idaho Criminal Rule 18.1," https://isc.idaho.gov/icr18-1; see also Dan Maurer, "Military Mediation as Military Justice?: Conjectures on Repairing Unit Cohesion in the Wake of Relational Misconduct," *Ohio State Journal on Dispute Resolution*, 28 (2013), 433–38.

96 10 U.S.C. §853a, and *Manual for Courts-Martial*, appendix 2.1 (nonbinding disposition guidance), pp. A2.1–2, at para. 3.2a.–m.

97 ABA, *Standards for the Prosecution Function*, 29, Standard 3–4.4.

98 *Law of War Manual*, p. 355 (para. 6.5.10); Department of Defense, *DoD Executive Agent for Non-Lethal Weapons (NLW), and NLW Policy*, Glossary, Directive 3000.03E (Washington, DC: Department of Defense, 2013).

99 "Investigations and Police: Overview of the Fourth Amendment," *Georgetown Law Journal Annual Review of Criminal Procedure*, 45 (2016), 250–69.

100 See, for example, United States v. Stellato, 74 M.J. 473 (Court of Appeals for the Armed Forces [C.A.A.F.] 2015).

101 Army Regulation 27–26, p. 69 (Rule 3.8(g) and (h)); and American Bar Association (ABA), 'Rule 3.8: Special Responsibilities of a Prosecutor," *Model Rules of Professional Conduct*, para. (g) and (h). https://bit.ly/3QBOAQI.

102 Neil Gordon, "Misconduct and Punishment," Center for Public Integrity, June 26, 2003. https://bit.ly/3dtPody; Supreme Court of the State of Illinois, *Illinois Rules of Professional Conduct*, Rule 8.4 and Rule 8.5 (2010), https://bit.ly/3JUmWfu.

103 Michael N. Schmitt, "Investigating Violations of International Law in Armed Conflict," *Harvard National Security Journal*, 2, no. 1 (2011), 31.

publish and discuss incidents that gave rise to violations, and train troops to avoid them in similar circumstances;[104] to have procedures and rules in place to sufficiently investigate[105] and punish[106] those offenders; and to offer compensation to the appropriate victims.[107]

Conclusion

These seven values provide the necessary scaffolding for designing and recommending contextually adequate and appropriate ROE or Rules for the Use of Force (RUF) in a *post bellum* environment – a *post bellum* Lieber Code of sorts. The intent of this code includes conditioning forces to not inappropriately turn to standard ROE intended for *jus in bello* or create a reactionary, *ad hoc*, unprincipled ROE, thereby creating risk or stressors on *post bellum* reconciliation, recovery, and return efforts. Again, the set of appropriate standards and rules would reduce the risk of breaching legal standards, outright prohibitions, and accepted norms, without imposing additional costs in the form of increased tactical risk to a commander's forces or mission. In this sense, ROE or RUF intended for *jus post bellum*, derived from these seven values, begin to look a lot like what animates the legal philosophy of what prosecutors would call "rehabilitative" or "restorative justice."[108]

We can define "restorative justice" as an alternative model or prosecutorial paradigm for certain kinds of "relational misconduct" that takes as relevant input a good deal more than merely the nature of the crime, identity of the victim, and harm caused as paramount considerations.[109] Rather, restorative justice is the deliberate, restrained use of criminal law and procedure as venues and methods recognizing that "neither the crime nor the parties can be detangled from the web of social interactions or community in which they play out."[110] Where Philpott argues that "restorative justice"

[104] *Law of War Manual*, p. 1082 (para. 18.6.2); Department of Defense, *DoD Law of War Program*, Directive 2311.01E (Washington, DC: Department of Defense, 2006) para. 5.11.9; see also W. Hays Parks, "The United States Military and the Law of War: Inculcating an Ethos," *Social Research*, 69, no. 4 (2002), 982–83.

[105] *DoD Law of War Program*, passim.

[106] *Law of War Manual*, p. 1087 (para. 18.7.3).

[107] Ibid., pp. 1107–9 (para. 18.16).

[108] Zvi D. Gabbay, "Justifying Restorative Justice: A Theoretical Justification for the Use of Restorative Justice Practices," *Journal of Dispute Resolution*, 2 (2005), 349; Evan Seamone, "Reclaiming the Rehabilitative Ethic in Military Justice: The Suspended Punitive Discharge as a Method to Treat Military Offenders with PTSD and TBI and Reduce Recidivism," *Military Law Review*, 208, no. 1 (Summer 2011), 1.

[109] Gabbay, "Justifying Restorative Justice," 349.

[110] Maurer, "Military Mediation as Military Justice?" 435.

on the scale of the nation-state is a critical component of political reconciliation, I offer it as analogous to the tactical, military scale.

Such a set of principles, if it existed, would serve several functions. It would regulate and potentially check armed forces' use of armed force, through more nuanced and contextually appropriate ROE. It would provide, in effect, a training guide for operating in ambiguous conflict settings where there is no longer a visibly discernible declared hostile force. It would establish a baseline of context-specific norms, and as Walzer noted,[111] a "vocabulary" from which external observers and stakeholders might evaluate, describe, condemn, or approve the use of armed force after the fact.[112] And it would, as Schuck wrote in 1994,[113] reveal what the soldier's "true attitudes" are under the circumstances, giving more meaning to their earlier *jus ad bellum* and jus in *bello* arguments.

Table of Cases

1. Brady v. Maryland, 373 U.S. 83 (1963) 2. Ex parte Quirin, 317 U.S. 1, 31 (1942)
2. Hamdan v. Rumsfeld, 548 U.S. 557 (2006)
3. Orloff v. Willoughby, 345 U.S. 83, 93–94 (1953)
4. Parker v. Levy, 417 U.S. 733 (1974)
5. Prosecutor v. Tadić, Case No. IT-94-1-T (Int'l Crim. Trib. for the Former Yugoslavia Aug. 10, 1995)
6. United States v. Barry, 78 M.J. 70 (C.A.A.F. 2018)
7. United States v. Boyce, 76 M.J. 242 (C.A.A.F. 2017)
8. United States v. Dancy, 38 M.J. 1 (C.M.A. 1993)
9. United States v. Howe, 37 M.J. 1062 (N-M Ct. of Crim. App. Review 1993)
10. United States v. Katz 389 U.S. 347 (1967)
11. United States v. Meek, 44 M.J. 1 (C.M.A. 1996)
12. United States v. Stellato, 74 M.J. 473 (C.A.A.F. 2015)
13. Youngstown Sheet & Tube Co. v. Sawyer, 343 U.S. 579 (1952)

111 Walzer, *Just and Unjust Wars*, p. xxv.
112 Ibid., 15, 20.
113 Schuck, "When the Shooting Stops," 984.

CHAPTER 10

Reconciliation Is Justice – and a Strategy for Military Victory

Daniel Philpott

St. Augustine and St. Thomas Aquinas, whom we may regard as the founders of the just war tradition, each held that the purpose of a just war is to secure a just peace. A just war must aim not simply at one side's victory but rather at a peace that entails justice between the combatant parties as well as a broad condition of justice within the societies involved in the war. Justice and peace, we might add, are connected in practice. Regnant injustice will become a source of instability and beget further violence and conflict.

Securing a just peace was not a problem unique to Augustine's or Aquinas's time. It is arguably the most difficult problem that the United States has faced in its foreign policy since the end of the Cold War, a far more difficult problem than securing formal military victory. Afghanistan and Iraq are quintessential instances. Following the attacks of September 11, 2001, the United States and its close allies quickly overthrew the government of the Taliban, which had harbored the attackers – and then stayed on to fight a civil war against the Taliban and its allies that continued through August 2021 when the United States withdrew. Likewise, the United States and a coalition of allies overthrew Iraqi president Saddam Hussein with relative ease in spring 2003 but became immersed in a complex civil war over the next several years and withdrew most of its troops in 2011, only to witness the rise of ISIS, or the Islamic State, in 2014, and become involved in defeating ISIS.[1] In 2011, U.S. President Barack Obama joined a NATO coalition in militarily overthrowing President Muammar al-Qaddafi of Libya, launching a succession of civil wars that wrought thousands of deaths and tens of thousands of refugees and were not concluded until fall 2020, and even then

[1] ISIS stands for Islamic State of Iraq and Syria. The group is also known as ISIL (Islamic State of Iraq and the Levant) and IS (the Islamic State). In this chapter, I will use the common acronym ISIS, though other usages appear in quotes.

uncertainly. While the United States was not involved in these wars, it bore moral responsibility for having ignited them. In other venues, too, including Somalia, Bosnia, and Kosovo, the United States intervened militarily and left behind dilemmas of peacebuilding.

What norms, practices, and guidance does ethics offer for the problems of peacebuilding? Despite Augustine's and Aquinas's insistence on a just peace as the proper telos of the use of force, the body of ethics that they launched – the just war theory – has offered far more moral guidance for undertaking and fighting war than it has for building peace in the aftermath of war. In recent years, some scholars have proposed a *jus post bellum* to accompany the traditional just war categories of *jus ad bellum*, which consists of norms for going to war, and *jus in bello*, which consists of norms for fighting war.[2] Most of these scholars focus on the obligations of intervening powers: What must they secure before they depart? To be sure, this has been a central question for U.S. policymakers, yet it is a narrow one. The question of departure is tied tightly to the question of securing a just peace, which is in turn tied to a larger set of questions regarding what this peace consists of. Should former combatants be involved in a new government? Is amnesty warranted for them? What about those who have committed war crimes? Should they or others face trials and punishment? How ought ethnic or religious minorities be included in a government? What is the best way to acknowledge and remember victims of violence? Should victims receive reparations? What is the role of apologies and forgiveness, if any?

These and other questions serve to reframe the issue as not simply one of exit strategy but rather as the more expansive problem of dealing with the wrongs and wounds of the past. This problem is not ethereal or hypothetical but is rather one that societies have debated all over the world in the last half century as they have confronted legacies of dictatorship, war, and genocide through efforts that have involved truth commissions, trials, reparations, apologies, and other practices. The issue for ethics may be summarized as: What is the meaning of justice in the aftermath of massive injustice?

A policymaker might object to such a capacious framing. The challenge of fighting an insurgency and promoting stability is already massive without trying to reconcile a populace. To be sure, nothing in my argument

[2] See Gary J. Bass, "Jus Post Bellum," *Philosophy and Public Affairs*, 32, no. 4 (2004), 384–412; and Stefano Recchia, "Just and Unjust Postwar Reconstruction: How Much External Interference Can Be Justified?," *Ethics & International Affairs*, 23, no. 2 (2009), 165–87.

here deigns to direct the soldier and their commanders in how to carry out a campaign or to judge when a war is unwinnable. Still, the operational problem of securing peace and stability may be far more dependent on larger questions of justice than the skeptical pragmatist wishes to admit. The problems that beset Iraq, Afghanistan, and Libya are much like those that South Africa, Rwanda, Germany, and Chile confronted through their efforts to deal with their pasts. It is not the role of military forces alone, or even primarily, to carry out the wide tasks of justice. Military leaders would be wise, though, to coordinate their efforts at stabilization with the efforts of international institutions, nongovernmental organizations (NGOs), judges, and eminent persons within societies, including religious leaders, to address past injustices.

The task of this chapter is to outline the contours of an ethic of peacebuilding, one that I developed in my 2012 book, *Just and Unjust Peace: An Ethic of Reconciliation*.[3] The core idea of the ethic is reconciliation. The roots of the ethic lie in religious traditions, particularly in Christianity, Judaism, and Islam, though its core commitments are articulable in secular language. Reconciliation is distinct from the ethic that dominates the international community, which I will call the liberal peace. The most important quality that distinguishes an ethic of reconciliation is that it offers a holistic restoration of persons and relationships in the aftermath of large-scale political injustice.

After setting forth the ethic of political reconciliation, I will then deploy it to Iraq, a site where peacebuilding dilemmas have been at their most difficult. Reconciliation indeed rose to prominence in the parlance of policymakers and other influential voices in the U.S., allied countries, and Iraq during the civil war fought in Iraq beginning in 2006. After the U.S. mostly withdrew in 2011, ISIS emerged and by summer 2014 had captured a substantial swath of Iraq, taking advantage of weakness in governance that was due in no small part to a lack of reconciliation. Today, now that ISIS is defeated in its governance of Iraq (though not at all eliminated),

[3] Daniel Philpott, *Just and Unjust Peace: An Ethic of Political Reconciliation* (New York: Oxford University Press). Other theorists of peacebuilding and transitional justice who give prominence to reconciliation include John Paul Lederach, *Building Peace: Sustainable Reconciliation in Divided Societies* (Washington, DC: United States Institute of Peace, 1997); Andrew Schaap, *Political Reconciliation* (London: Routledge, 2004); Ernesto Verdeja, *Unchopping a Tree: Reconciliation in the Aftermath of Political Violence* (Philadelphia, PA: Temple University Press, 2009); Colleen Murphy, *A Moral Theory of Political Reconciliation* (Cambridge: Cambridge University Press, 2010); Colleen Murphy, *The Conceptual Foundations of Transitional Justice* (Cambridge: Cambridge University Press, 2017); and Catherine Lu, *Justice and Reconciliation in World Politics* (Cambridge: Cambridge University Press, 2017).

the problems of communal division remain. U.S. military troops, which helped to defeat ISIS, will likely remain, too. The U.S. interest in mitigating these divisions is therefore strong.

The Liberal Peace

The past half century might be called an "age of peacebuilding." This is not to be confused with an age of peace, a characterization quickly refuted by massive bloodshed in Rwanda, Yugoslavia, Timor Leste, South Sudan, and the Democratic Republic of Congo. The age of peacebuilding is rather a wave of institutions and political practices through which countries have confronted the past injustices of dictatorship, civil war, and genocide in hopes of laying the foundations of a peace settlement or a nascent democracy – much the outcome the United States has hoped for in Afghanistan and Iraq. More than forty truth commissions have taken place since the mid-1970s. Reviving the precedent of the Nuremberg trials, two international tribunals for Rwanda and Yugoslavia were created in the middle 1990s and were succeeded by a permanent International Criminal Court, which began operation in 2002. Reparations settlements have multiplied and proliferated, as have political apologies, which spiked in the 1990s. The discourse and practice of forgiveness, which had scant pedigree in the modern nation-state or political thought, emerged in locales such as South Africa, Uganda, and Northern Ireland. Localized community justice forums have taken place in countries such as Rwanda and Timor Leste. Election monitoring and U.N. peacekeeping and peacebuilding exercises have proliferated. By means of global communication and travel, through imitation and with incremental improvements, these institutions and practices have spread rapidly. Though with varying levels of success, these efforts have taken place in dozens of countries.

Taken together, these activities have come to be known as transitional justice.[4] Of what does justice consist, though, in the eyes of those who lead, participate in, and advocate for such efforts? Among officials in international organizations, international lawyers, human rights activists, and Western NGOs – the most prominent in this field being the International Center for Transitional Justice (ICTJ) – the predominant answer to this question is what may be called the "liberal peace." The pedigree of this orthodoxy is

[4] Paige Arthur, "How 'Transitions' Reshaped Human Rights: A Conceptual History of Transitional Justice," *Human Rights Quarterly*, 31, no. 2 (2009), 321–67.

the Western Enlightenment, especially philosophers such as John Locke and Immanuel Kant, whose notions of justice derive from a far older conception rooted in Roman law. Seven tenets describe the liberal peace:

First, armed conflict and dictatorships must come to a halt through a peace settlement or a transition to a constitutional liberal democracy.

Second, this peace settlement must be based on the principles of human rights, democracy, the rule of law, and free market economics.

Third, the perpetrators of past crimes, especially the most egregious ones – those which international law calls war crimes, crimes against humanity, and genocide – should be tried fairly in a court of law, and, if found guilty, punished. Relatedly, "vetting" practices may be applied to large bodies of people who were employees or agents of dictatorships but who cannot be tried realistically on account of their numbers, as took place in the Czech Republic, Iraq, and other countries.[5]

Fourth, victims have the right to reparations for their suffering. For injustices on a mass scale, reparations are determined and distributed through general policies aimed at large populations rather than through individual court cases.

Fifth, the liberal peace, consistent with liberal thought, proffers a public-private distinction and is thus wary of the pursuit in the political realm of personal and spiritual healing, the transformation of emotions, and interpersonal reconciliation, all of which ought to be relegated to the private realm.

Sixth, the liberal peace is secular. Reflecting the commitment of liberal philosophers to public reason, it holds that public discourse in matters of transitional justice ought to be expressed in secular rationales, not religious ones.

Seventh, in the thinking of the liberal peace, the leading actors in transitional justice are governments, international organizations, international lawyers, and NGOs that promote transitional justice or certain aspects of it like reparations or human rights. It is because the commitments of the liberal peace are espoused and promoted by these powerful and prestigious actors that it stands as the global "orthodoxy" of transitional justice.[6]

[5] This tenet arguably holds the most prestige among proponents of the liberal peace. Were the liberal peace a religion, its cathedral would be the twin glass towers of the International Criminal Court that stand in the Hague. Its mortal sin would be "impunity," the criticism that the human rights community has voiced against blanket amnesties.

[6] Authoritative statements of the liberal peace can be found in United Nations documents. See United Nations Secretary General, "Supplement to an Agenda for Peace" (1995); and Lakhdar Brahimi, "Report of the Panel on United Nations Peace Operations" (United Nations, 2000). See also Arthur, "A Conceptual History of Transitional Justice."

An alternative paradigm for addressing the aftermath of massive violence and injustice, though, has also sprung up across the globe in recent decades: reconciliation. Sociologist Jonathan Van Antwerpen calls reconciliation the global "heterodoxy" for transitional justice, implying that it is widely shared yet stands as an alternative to the liberal peace.[7] Reconciliation has been strongly associated with religion, though not exclusively. In my account, I root it in Judaism, Christianity, and Islam, among which reconciliation's core tenets can be the object of an "overlapping consensus."[8] Potentially, this same consensus could be extended to other religious traditions as well.

The religious dimension is sometimes controversial among proponents of the liberal peace. For at least two reasons, though, the role of religion merits consideration.

First, the involvement of religious leaders on the ground in these issues is far greater than the official, secular, discourse would suggest. Countries that have emerged from war and dictatorship have been concentrated in Eastern Europe, Africa, Latin America, and East Asia, where, with the exception of Eastern Europe, religion and politics are far less separated than they are in developed democracies. Not surprisingly, then, we see Archbishop Tutu leading the Truth and Reconciliation Commission in South Africa, the Catholic Church conducting its own truth commission in Guatemala, Muslims expressing favor for forgiveness and reparations in Iraq, and Pentecostal pastors injecting forgiveness and reconciliation into the Gacaca courts in Rwanda. In all of these cases, transitional justice provides a very different character than the human rights community had envisioned.

Second, religions carry teachings and practices that hold promise for building a just peace in the wake of massive violence and injustice. These teachings and practices are most widely and commonly expressed through the concept of reconciliation.

I also contend, though, that in settings of religious pluralism or in ones where secular language is official – say, the International Criminal Court or a domestic amnesty hearing – religious actors ought to search for rationales that are shared with other traditions and languages. Here, the

[7] Jonathan Van Antwerpen, "Reconciliation as Heterodoxy," in Jennifer Llewellyn and Daniel Philpott (eds.), *Restorative Justice, Reconciliation, and Peacebuilding* (New York: Oxford University Press, 2014), 77–117.

[8] On overlapping consensus in justifications for an ethic of political reconciliation, see Philpott, *Just and Unjust Peace*, pp. 18–22.

political ethic of reconciliation benefits from finding analogous principles in secular discourse. For instance, much of the ethic finds resonance with the school of thought known as restorative justice, which is often expressed in secular terms.[9]

What, then, are the core ideas that religious traditions bring to an ethic of political reconciliation? At the center of the ethic is a notion of justice that is the virtual equivalent of reconciliation. That reconciliation would be justice will sound strange to citizens of modern liberal democracies steeped in thinking of justice as rendering another her due. This more familiar way of thinking about justice translates into rights, fairness, equality, and punishment justly meted out – core ideas of the liberal peace. Justice in the Bible, the Quran, and the Jewish, Christian, and Islamic traditions, by contrast, means something far wider that can be called the justice of right relationship. Expressed by the Hebrew *sedeq* (or its feminine form, *sedeqah*), the Greek *dikaiosune*, and the Arabic *'adl*, such justice is the entirety of human obligations toward other persons and God. This justice may well encompass rights, retribution, and the like, but it also involves duties that do not correspond neatly to rights and dimensions of addressing past wrongs that do not correspond neatly to retribution. That is, it involves dimensions of relationality and human flourishing that exceed the boundaries of the justice of rendering due. This wide justice might be conceived of as a comprehensive state of right relationship or as a process of restoring right relationship after it has been sundered by violence and injustice. Since reconciliation also means the restoration of right relationship, reconciliation converges with justice.[10]

Reconciliation, conceived as restoration of right relationship, also finds expression in the three religious traditions at hand. Reconciliation is most explicit in Christianity, where it appears fifteen times in the Letters of the Apostle Paul, denoting God's restoration of the world to himself and the consequent restoration of relationship that Christians are called to enact in response.[11] Reconciliation is not found explicitly in the Old Testament, or the Hebrew Bible, but finds equivalent expression in certain uses of *sedeq* that denote God's salvation of the people of Israel after they have

[9] My approach here draws from that of Christopher J. Eberle, *Religious Conviction in Liberal Politics* (New York: Cambridge University Press, 2002). Eberle argues against "public reason," which holds that political discourse ought to refrain from religious or other "comprehensive conceptions," but calls for religious people to find common principles with people of other traditions and convictions.

[10] I lay out the argument that this is how these religious traditions conceive of justice in Philpott, *Just and Unjust Peace*, 122–5, 135–8, 152–6.

[11] See, for instance, 2 Corinthians 5:14–21.

sinned and suffered punishment. In the Islamic tradition, reconciliation is expressed through the Arabic *sulh*, which means the settlement of differences through a restoration of relationship, and *musalaha*, which translates directly to reconciliation and connotes comprehensiveness. Today, *sulh* and *musalaha* denote restorative practices in communal settings and are viewed by Muslims as resources for peacebuilding.[12]

In these traditions, reconciliation not only describes justice but is also a vision of peace. The Hebrew word for peace, *shalom*, appears commonly in the Bible and means something much like holistic right relationship, involving health, prosperity, economic justice, political justice, honesty, and moral integrity among the members of the community. The New Testament word for peace, *eirene*, is the direct Greek translation of *shalom*, and carries its comprehensive and communal meaning. *Salam*, the Arabic word for peace found in the Quran also means a general state of right relationship, much like the Jewish *shalom*. In all three traditions, peace, like justice, is a state of affairs where people not only rightly assert certain actions that are due to them but also realize a wider set of goods and dimensions of right relationship.[13]

The virtue that animates the process of restoring right relationship is mercy. In all of the religious traditions, mercy involves a wide restoration of what is broken, in contrast to the far narrower notion of mercy in the Enlightenment tradition, which is a waiving of deserved punishment. In the Bible and in the Quran, mercy is a general will to restore right relationship with respect to the wide range of ways that it can be broken. Such a mercy does not sharply contrast with but rather strongly resembles the biblical concept of saving justice, and hence, reconciliation. Likewise, Arab words for mercy such as *rahamin* connote a general willingness to restore. Mercy, then, is not at odds with justice but is rather a dimension of the justice that restores relationships.[14]

How might reconciliation come to be practiced in modern politics? An answer begins with identifying the range of ways that violence and injustice ruptures right relationship, which I call the wounds of injustice.

There are at least six of these wounds:

First, *the violation of the victim's basic human rights.*

[12] Mohammed Abu-Nimer, *Nonviolence and Peace Building in Islam* (Gainesville: University Press of Florida, 2003), pp. 102–8.

[13] For my understanding of reconciliation in these traditions, see Philpott, *Just and Unjust Peace*, pp. 122–25, 135–38, 141–50, 152–56.

[14] For my understanding of mercy in these traditions, see ibid., pp. 126–27, 138–41, 156–57.

Second, *the many forms of harm to the victim's person*, including death, the death of loved ones, permanent bodily injury, grief, humiliation, trauma, the loss of wealth and livelihood, the defilement of people's race, ethnicity, religion, nationality, or gender, sexual violation, and other harms.

Third, *victims' ignorance of the source and circumstances of the political injustices that harmed them.*

Fourth, *the failure of members of the community to acknowledge victims' suffering, either through ignorance or indifference.*

Fifth, *the "standing victory" of the perpetrator's political injustice.* Violence and other political injustices leave behind not only harms to the victim's person but also an unchallenged, undefeated message of disregard for the victim's dignity.

Sixth, *the wounded soul of the perpetrator.* Deep in several religious and philosophical traditions is the idea that evil injures the soul of the perpetrator. Often, this injury will redound in severe psychological damage.

Insofar as these six forms of wounds are inflicted directly by political injustices, they may be called "primary wounds." Primary wounds lead further to "secondary wounds" – new acts of injustice and withdrawals of assent from new regimes that arise from the emotions of fear, hatred, resentment, and revenge that emanate from memories of the original injustices themselves. As countries like Bosnia, Ireland, and Rwanda attest, these secondary wounds further stunt the project of building just and stable political orders, sometimes for generations.[15]

Correspondingly, an ethic of political reconciliation aspires to restore persons and relationships with respect to the range of primary and secondary wounds that political injustices inflict. It is through practices that the ethic is enacted. There are six of these as well, though they do not correspond in one-to-one fashion with the wounds. They are: building institutions for social justice, acknowledgment, reparations, punishment, apology, and forgiveness. Each manifests reconciliation's core logic of restoration of right relationship and restores persons and relationships with respect to distinct dimensions of woundedness. Inasmuch as these practices heal primary wounds, they bring about what can be called "primary restorations." These restorations can then elicit "secondary restorations" that take the form of "social capital," including an increase in popular trust in the political order, democratic participation, and identification with the nation.[16]

[15] On wounds in an ethic of political reconciliation, see ibid., pp. 30–47.
[16] On primary restorations and secondary restorations, see ibid., pp. 56–58.

As the age of peacebuilding has evidenced, these restorative practices will always be achieved only in part, hampered by power and unresolved differences over the meaning of justice, and limited by their sheer size and complexity. But as the past generation has also shown, the practices are not merely the brainchild of armchair theorists. Each of the six has in fact taken place in numerous settings over the past generation, however messily, and might now contribute something to the building of peace in Iraq.

Reconciliation in Iraq

The six practices manifest the restorative logic of an ethic of political reconciliation. In Iraq, this ethic implies a wider set of activities than what policymakers now mean by political reconciliation, but activities that may well be vital to building a peace that lasts.

The formidable task of political reconciliation in Iraq is conveyed by the story of a Shiite Iraqi, Sadri Adab Diwan, which appeared in the *New York Times* in 2003. Saddam Hussein's regime had abducted and killed Diwan's younger sister, Hamaa, in 1980 because she had given a Koran to a classmate. Saddam Hussein's regime not only favored Sunnis but also was secularist, especially during its early years. The violence that emerged after Hussein's fall is explained well by Diwan's vow: "If I catch Saddam, I won't kill him. That won't be enough. I'll suck his blood. And if he escapes, I'll follow him to the ends of the earth."[17] Not only Shias, but also Kurds, who were also treated brutally by Hussein, pursued revenge. Sunnis retaliated. Sunni terrorists destroyed the Shiite Golden Mosque of Samarra in February 2006, sparking a full-fledged civil war. After the civil war died down around 2008, the same loyalties, fear, and emotions led Shiites to exclude Sunnis from government and helped to create the power vacuum that led to ISIS. To this day, moderating and assuaging these sentiments is indispensable to any achievement of stability in Iraq.

Reconciliation was in fact on the minds and in the language of policymakers, not only in the United States, but also in Europe and Iraq, during this period. Flummoxed by the civil war, President George W. Bush formed a bipartisan Iraq Study Group, whose report of November 2006 mentioned reconciliation sixty three times and devoted a major section to it.[18] U.S. officials such as Secretary of State Condoleezza Rice, U.S.

[17] Susan Sachs, "Iraqis Seek Justice, Or Vengeance, for Victims of the Killing Fields," *The New York Times*, November 4, 2003.

[18] The Iraq Study Group, *The Iraq Study Group Report* (New York: Random House, 2006).

Ambassador Zalmay Khalilzad, and his successor, Ryan Crocker, voiced the term frequently, as did U.N. Secretary-General Kofi Annan, Special Envoy Lakhdar Brahimi, and heads of state such as U.K. Prime Minister Tony Blair, and Italian Prime Minister Silvio Berlusconi. Iraq's Prime Minister Nouri Al-Maliki issued a 28- point plan for reconciliation in June 2006, one reputed to be influenced by South Africa's Truth and Reconciliation Commission.[19] Reconciliation was on the lips of Iraqi civil society leaders, including religious ones, too. Perhaps most consequentially, U.S. General David Petraeus made reconciliation an orienting concept of his "surge" strategy that was largely responsible for bringing the civil war to a close and paving the way for the departure of U.S. troops from Iraq.[20]

What did political officials mean, though, when they spoke of reconciliation? Generally, what they had in mind was incorporating Sunnis into the government; reversing draconian de-Ba'athification policies, which had excluded a vast swath of government officials under Hussein from serving in a future government; distributing oil revenue among Sunnis, Shiites, and Kurds; holding provincial elections that would strengthen local autonomy; widening amnesty for members of Saddam's security forces as well as for insurgent fighters; and granting greater autonomy to the Kurdish people in the north.

Such political goals were doubtless compatible with an ethic of political reconciliation, but were they sufficient for securing peace and stability in Iraq, the primary foreign policy goal of the U.S., its allies, and the Iraqi government? Iraq's subsequent trajectory leaves room for doubt. Arguably what was needed was a wider array of practices that addressed the memories and emotions that fueled the desire for revenge that brought about the civil war and contributed to the rise of ISIS. This insight lies behind the truth commissions that over forty countries have undertaken during the age of peacebuilding, at least some of which contributed to the stability of nascent regimes and peace settlements. A stable Iraq requires stable institutions, which in turn require legitimacy in the eyes of the population, which further requires dealing with the wounds that have undermined this legitimacy.

How might the practices that enact the ethic of political reconciliation that I have outlined approach the task in Iraq? The first practice, building socially just institutions, is the one that most converges with the liberal peace as well as with the goals of U.S. policymakers. It involves building

[19] *The Irish Times*, "Iraqi Prime Minister Unveils Reconciliation Plan," June 26, 2006, https://bit.ly/3Pn8iP5.

[20] Peter R. Mansoor, *Surge* (New Haven, CT: Yale University Press, 2013).

stable institutions based upon a constitution that promotes human rights, democratic governance, and the rule of law. The wounds that this practice addresses are the colossal violations of these principles under the regime of Saddam Hussein, during the civil war, and at the hands of ISIS.

Today, the major shortcoming in this practice is the failure of the majority Shia community to include the Sunni minority in governance. Since Hussein's fall, Shia leaders have used their positions to favor the Shia majority, which Hussein had brutally excluded from power. In recent years, however, a popular backlash against such sectarianism has sprung up, creating a demand for sharing of power that would not exclude people from participation in government on account of their religious affiliation.[21]

A related proposal, one that has received widespread support, including from former Prime Minister Haider Al-Abadi (who held office from 2014 to 2018), is to devolve authority to Iraq's provinces so as to allow Sunni majority regions greater autonomy. Some Shia elites oppose the idea, fearing that decentralized security structures in particular will destabilize the country. In response, the U.S. Agency for International Development has encouraged the Iraqi leadership to pursue a viable plan for decentralization through what is known as the Governance Strengthening Project.[22]

Critical for alleviating sectarian divisions is also religious freedom, a principle that is ensconced in the human rights conventions and in the constitutions of the vast majority of the world's states. The freedom of persons and communities to practice their religious faith without coercion is a critical source of harmony in a society riven by sectarian divisions.[23] Religious freedom is strongly affirmed by the Catholic Church, the major Protestant Churches, and prominent Jewish organizations and intellectuals. It is not as widely affirmed among Muslim scholars and clerics, though the tradition contains teachings and historical precedents that support the principle and help it to grow more widely.[24]

Consistent with the claim of an ethic of political reconciliation that law, rights, and constitutions are not sufficient for addressing the wounds of the past, other restorative practices are demanded as well, the second of

[21] Ranj Alaaldin, "Sectarianism, Governance, and Iraq's Future," Brookings Institution Doha Center, analysis paper no. 24 (2018), https://brook.gs/3Qpuxp5.

[22] Ibid.; and Todd Diamond, *The Role of Decentralization in Combatting Extremist Influence in Iraq* (The Atlantic Council, 2017), https://bit.ly/3cPoZFx.

[23] See Daniel Philpott, "Religious Freedom and Peacebuilding: May I Introduce You Two?" *The Review of Faith & International Affairs*, 11 (2013), 31–37.

[24] See Daniel Philpott, *Religious Freedom in Islam: The Fate of a Universal Human Right in the Muslim World Today* (New York: Oxford University Press, 2019), pp. 177–205.

which is acknowledgment. Enacted by the surrounding society under the imprimatur of its government, acknowledgment confers recognition on the victims of human rights violations and their suffering. The wound that acknowledgment addresses is the lack of recognition and of knowledge of injustices on the part of the society. While knowledge and recognition are intrinsically important – "primary restorations" – the political practice of acknowledgment also contributes to the building of stable regimes and peace settlements – "secondary restorations" – by leading victims to forego revenge and lend their loyalty to new orders.

Acknowledgment takes many forms, including museums, monuments, days of remembrance, ceremonies, and other forms of remembrance. The most important, innovative, and widely discussed forum for acknowledgment is the truth commission. All truth commissions are charged by a government with compiling a public record of past injustices and are granted a finite amount of time to look into the events of a prescribed period in history. A truth commission's practice of acknowledgment is generally at its best when the recognition that it confers is personal, meaning that a victim testifies to an audience of empathetic hearers, which may take place in public hearings and other form of witnessed testimony.

To date, no truth commission has taken place in Iraq, although respected authorities in the international community, as well as the United States Institute of Peace (USIP) and the ICTJ have proposed that one take place.[25] A major obstacle is Iraq's weak government institutions, which hinder the emergence of a strong, well- financed, and impartial commission. Other countries such as Sierra Leone and El Salvador have overcome this problem, though, by adopting truth commissions that combine domestic personnel and authority with those of an international body such as the United Nations. Truth commissions and similar forums can also take place on the regional and community level. In 2004, the ICTJ and the Human Rights Center of the University of California, Berkeley published *Iraqi Voices*, an extensive survey of Iraqis that evidenced broad favor for truth-seeking processes, especially those conducted through local forums and by NGOs.[26] In the past decade, both Iraqi leaders and international NGOs have promoted localized reconciliation initiatives in Iraq that are designed to confront the truth about past injustices. In 2017, the United Nations Development

[25] Joe Sterling, "In Iraq, Truth Commission Idea Gains Traction," *CNN*, March 25, 2009, https://cnn.it/3JTB4WH.

[26] Phuong N. Pham et al., *Iraqi Voices: Attitudes Toward Transitional Justice and Social Reconstruction* (Berkeley: The International Center for Transitional Justice and the Human Rights Center at the University of California, 2004).

Programme (UNDP), in collaboration with other international actors, created a project, Support for Integrated Reconciliation in Iraq, one of whose key activities is the establishment of Local Peace Communities that gather and publicize testimonies of human rights violations.[27] Proposals for acknowledgment have also envisioned documentation centers, museums, memorials, and the unearthing of mass graves.

The third practice, reparations, may be thought of as acknowledgment materially fortified. Like acknowledgment, reparations confer recognition upon victims, but they are distinguished by their material dimension. In the case of mass human rights violations, it is usually the state that confers reparations, but sometimes perpetrators do as well. Reparations address the wound of the lack of recognition through their symbolic communication but they also address the victim's loss of property, health, and other goods by providing money, health services, and other forms of benefit. Reparations also resemble acknowledgment in furthering secondary restorations such as reducing the resentment of victims and securing their assent to a new regime or settlement.

The report, *Iraqi Voices,* showed strong and widespread support among Iraqis for material and symbolic compensation for victims who suffered human rights violations. Consistent with this support, the Iraqi parliament passed a law in 2009 titled "Law No. 20: On the Compensation for Victims of Military Operations, Military Mistakes and Terrorist Actions," which has formed the primary framework for reparations in Iraq. The law addresses injustices that took place following the U.S.-led invasion of Iraq in 2003 and provides reparations to people who have suffered a wide range of injuries ranging from the death of loved ones to the loss of employment and schooling. As of 2016, the government had processed 65,046 cases involving property damage and 118,894 cases involving death, injury, or the loss of persons, and had awarded 420 billion Iraqi dollars ($355 million) in reparations to victims of the Iraq War.[28] Today, Iraq faces the new challenge of compensating victims of ISIS, whose predations postdate the law. Human rights victims under Saddam Hussein's rule arguably deserve compensation as well. For reparations to contribute to a stable constitutional democracy in Iraq, they must be distributed widely and evenly across victim groups.

[27] United Nations Development Programme, "Fast Facts: Support for Integrated Reconciliation in Iraq," December 1, 2017, https://bit.ly/3QI6ph3.

[28] Clara Sandoval and Miriam Puttick, "Reparations for the Victims of Conflict in Iraq: Lessons Learned from Comparative Practice," Ceasefire Centre for Civilian Rights and Minority Rights Group International (2017), https://bit.ly/3zlzRCA.

The fourth practice, punishment, is one of the most contentious issues in paradigm debates over transitional justice, which have arrayed reconciliation against punishment, and mercy against justice. Punishment, though, may constitute a practice of reconciliation if it is justified along the lines of "restorative punishment," a concept developed by Christian theologian Christopher B. Marshall, who grounded it in biblical texts.[29] The Quran and Arabic tribal traditions contain rationales and actual practices of restorative punishment as well.[30] Restorative punishment is a variant of restorative justice, which has been implemented, mostly on the community level, in Western countries in recent years. The core idea of restorative punishment is that crimes do not merely violate a law but also wound persons and relationships, and that addressing these wounds and restoring these relationships is the purpose of punitive practices. The wounds that restorative punishment addresses are the "standing victory" of the perpetrator's act of injustice and the damage to the soul of the perpetrator herself that his injustice inflicted. With respect to large-scale injustices, restorative punishment serves to address these wrongs in a manner that also contributes to stability and peace – a secondary restoration.

In practice, restorative punishment might involve long-term imprisonment for architects of atrocity, those who plan and oversee large-scale human rights violations, as well as for perpetrators of politically motivated acts of murder, torture, and assault. Following the fall of Saddam Hussein, he and other top leaders of his regime were tried by the Iraqi High Tribunal, a court established under Iraqi law to try alleged perpetrators of crimes prohibited in international law, including genocide, war crimes, crimes against humanity, and other major crimes that were committed between 1968 and 2003, the years of Saddam's rule. It was the only tribunal for international law crimes in the twenty-first century that applied the death penalty. Manifestly unrestorative was the publicly televised hanging of Saddam Hussein in December 2006, which fueled conflict between Sunnis and Shiites both in Iraq and throughout the Arab world.[31]

Consistent with many other countries that have practiced transitional justice, lower-level perpetrators might receive amnesty, which restorative

[29] Christopher D. Marshall, *Beyond Retribution: A New Testament Vision for Justice, Crime and Punishment* (Grand Rapids, MI: Wm. B. Eerdmans Publishing, 2001).

[30] For a discussion of this and more broadly of restorative punishment in the context of political transitions, see Philpott, *Just and Unjust Peace*, pp. 219–50.

[31] See John D. Carlson, "Discerning Justice in the Trial and Execution of Saddam Hussein," in David K. Linnan (ed.), *Enemy Combatants, Terrorism, and Armed Conflict Law: A Guide to the Issues* (Westport, CT: Praeger Security International, 2008), 307–26.

punishment would sanction only if it is conditional upon their participation in forums such as truth commissions where they face their victims, express contrition, and/or consent to perform service or pay reparations.[32] Forums that combined these goals took place in Rwanda and Timor Leste.[33] Restorative punishment can also involve vetting, the debarring of perpetrators from holding public office or other positions, which took place through the de-Ba'athification policies imposed by the U.S.-dominated Coalitional Provisional Authority. The process was sweeping and involved low levels of scrutiny of alleged wrongdoers, yielding the criticism that it fell short of principled punishment. Few forums took place that were designed explicitly on the principle of restorative punishment.

In more recent years, Iraq has focused on bringing ISIS terrorists to justice and has done so in cooperation with the international community. The United Nations Security Council established the U.N. Investigative Team to Promote Accountability for Crimes Committed by Da'esh/Islamic State in Iraq and the Levant (UNITAD), headed by international lawyer Karim Khan, which is now investigating crimes and plans tribunals. In Khan's view, it is only through public trials, which expose citizens to evidence of ISIS's crimes, that the ideology of radical jihadism can be exposed and undermined. Moreover, public trials allow the views of ISIS to be distinguished from those of the larger Sunni community, which is crucial for reducing tensions between the Sunni and Shiite communities.[34]

The Iraqi government has also pursued punitive measures against ISIS insurgents. As of 2018, it had detained more than 19,000 people on terrorism charges related to ISIS activity, convicted over 3,000, and put some 250 to death.[35] Khan argues that these trials have produced numerous justified convictions, but that most of them have taken place in private and thus have failed to bring the victims of ISIS crimes face-to-face with the perpetrators.[36] Voices in the international community have also raised questions about due process and overly sweeping prosecutions.[37]

[32] On "restorative amnesty," see Louise Mallinder, "Amnesties in the Pursuit of Reconciliation, Peacebuilding, and Restorative Justice," in Jennifer Llewellyn and Daniel Philpott (eds.), *Restorative Justice, Reconciliation, and Peacebuilding* (New York: Oxford University Press, 2014), 138–73.

[33] Philpott, *Just and Unjust Peace*, pp. 223–30.

[34] Sarah Benhaida, "Head of UN Investigators Calls for an IS 'Nuremberg'," *Yahoo News*, July 29, 2019, https://yhoo.it/2OqPWCR.

[35] Mara Redlich Revkin, *The Limits of Punishment: Transitional Justice and Violent Extremism* (Tokyo: United Nations University, 2018), https://bit.ly/2zI6nQC; Qassim Abdul-Zahra and Susannah George, "Iraq Holding More Than 19,000 Because of ISIS Militant Ties," *Associated Press*, March 21, 2018, https://apnews.com/article/aeece6571de54f5dba3543d91deed381.

[36] Benhaida, "Head of UN."

[37] Revkin, *The Limits of Punishment*; Abdul-Zahra and George, "Iraq Holding More Than 19,000."

Since the fall of Saddam, then, the Iraqi government, sometimes in collaboration with the international community, has conducted extensive judicial trials of perpetrators of the most serious crimes. Restorative punishment welcomes accountability and favors such trials for the most serious crimes. However, deficiencies in the rule of law create the risk that such trials will be perceived as instruments of revenge. The kinds of forums that most exemplify restorative punishment, combining the testimony of victims in the presence of perpetrators and outcomes that serve to restore communities, have not materialized.[38]

The fifth practice, apology, has risen sharply in frequency in global politics since around 1990.[39] It is usually heads of state who voice the apologies, and they do so most often with respect to crimes committed under a previous head of state or regime. When a public apology is successful – widely applauded and well-received by victims – it can bring healing to a country riven by war or dictatorship, serving to nullify the "standing victory" of injustices, to delegitimize the previous regime, leader, or faction in a war who committed the injustices, and to contribute to the legitimacy of a new regime. In a December 2015 report on apologies as a form of transitional justice, the ICTJ wrote that "official public apologies are an important element of transitional justice policy."[40] The Federal Republic of Germany's apologies for the crimes of the Nazi era have exceeded the apologies of any other country and have been broadly successful.

Few, if any, public apologies have taken place in Iraq since Saddam Hussein's fall, either on the part of government officials, opposition leaders, religious leaders, or other public figures. One reason is noted by the ICTJ: "more often apologies come only after the passage of a significant length of time, sometimes over a number of generations, several transitions of governments, or after a succession of regimes." The ICTJ report also observes that apologies often arise from truth commissions, which have not taken place in Iraq.[41]

Forgiveness is the most distinctly theological of the six practices and the most rarely practiced in the political realm. It is also the practice of which proponents of the liberal peace are most skeptical, both of whether

[38] Revkin, *The Limits of Punishment*.

[39] Aaron Lazare, *On Apology* (New York: Oxford University Press, 2004), pp. 6–8; and Philpott, *Just and Unjust Peace*, pp. 198–200.

[40] Ruben Carranza, Cristián Correa, and Elena Naughton, *More than Words: Apologies as a Form of Reparation* (New York: International Center for Transitional Justice, 2015).

[41] Ibid.

it does take place and whether it ought to take place. To ask victims to forgive, they argue, is to re-burden them with the task of repair when it is they who should be receiving reparative measures like apologies, acknowledgment, and reparations. Of religious and philosophical traditions, it is Christianity that commends forgiveness most strongly and directly. The New Testament teaches that forgiveness is an imitation of and participation in God's redemptive action. Islam too teaches forgiveness. The Qur'an states that God forgives the repentant sinner and calls victims to forgive a wrongdoer who apologizes. In an ethic of political reconciliation, especially one rooted in religious traditions, forgiveness is viewed as restorative. To forgive is to forego retribution and resentment but also to will the moral restoration of the perpetrator and, insofar as it is possible, the restoration of right relationship. It can fortify victims in their agency by enabling them to be constructors of the good, defeat the standing victory of the perpetrator's injustice by condemning and then overcoming it, invite the perpetrator to repent, stunt cycles of revenge, and contribute to peace and stability in a new settlement or regime.

Few studies have charted how often or under what conditions forgiveness takes place in the aftermath of large-scale political injustices.[42] One exception is a study that I conducted in Uganda in partnership with the Refugee Law Project that assessed the attitudes and practice of 640 people across five regions that had experienced political violence. Contrary to the expected rarity of forgiveness in the liberal peace, solid majorities of respondents were favorable to forgiveness. Asked, "What would you like to see happen to members of rebel groups who committed human rights violations," 60 percent of respondents answered "yes" to the choice of "forgive"; 86 percent of respondents answered "agree" to the statement, "[i]t is good for victims to practice forgiveness in the aftermath of violence"; and 68 percent of respondents who had experienced violence answered "yes" to the question, "[d]id you personally forgive the perpetrator of the act of violence against you?"[43] Forgiveness is regarded far more favorably, and practiced far more commonly than the liberal peace would expect, at least in Uganda. Strong discourses of forgiveness and the practice of forgiveness, though undocumented in its extent, have also existed

[42] One exception is the work of political scientist David Backer. See David Backer, "Victims' Responses to the Truth Commissions: Evidence from South Africa," in Muna Ndulo (ed.), *Security, Reconciliation and Reconstruction: When Wars End* (London: University College of London Press, 2007), 165–96.

[43] Refugee Law Project, *Forgiveness: Unveiling an Asset for Peacebuilding* (Refugee Law Project, 2015), https://bit.ly/3JgN3wV.

in South Africa, Rwanda, Sierra Leone, Northern Ireland, Timor Leste, and other countries during the age of peacebuilding. What about Iraq? *Iraqi Voices* shows that ordinary Iraqis support forgiveness significantly. In 2018, Vatican Secretary of State Cardinal Pietro Parolin preached on forgiveness during a Christmas mass in the Syro-Catholic Cathedral near Mosul, Iraq, telling an audience of Christians who had been displaced by ISIS that forgiveness was a "tangible sign of the Christian faith" and could contribute to reconciliation in Iraq.[44] Little evidence exists regarding how widely Iraqis have practiced forgiveness, though.

Realism, Religion, and Reconciliation

The practices that make up an ethic of political reconciliation have taken place in Iraq unevenly and with mixed results. This has direct implications for the urgent moral and policy problem faced by the United States and its allies in Iraq and in many other locales since the end of the Cold War: How can a sustainable peace be secured in countries in which the U.S. and its allies have strong interests at stake but in which achieving this peace requires far more than an initial military victory? A more robust, holistic, and integrated strategy of reconciliation, I argue, could contribute to a far stronger peace in Iraq and to relieving the pressure on the United States and its allies to intervene again in the future.

As I argued above, though the task of reconciliation does not fall solely, or even primarily, on the military, the military ought to recognize the need for reconciliation and to coordinate its actions with others who are carrying out restorative practices. Again, the U.S. military made reconciliation a central strategy during Iraq's civil war of 2006–08, forming "Provincial Reconstruction Teams," "Concerned Local Citizens Programs," and bottom-up reconciliation programs that coordinated U.S. efforts with those of local Iraqi citizens. Still, the primary work of practices of reconciliation – a national truth commission, local reconciliation forums, reparations programs, apology, forgiveness, and restorative punishment – are best conducted by international institutions, NGOs, the national government in Iraq, and respected Iraqi civil society leaders.

In addition to the UNDP activity mentioned above, the USIP has also undertaken several initiatives to promote local reconciliation. Often, the USIP, in conjunction with Iraqi officials, acts as a mediator

[44] Christine Rousselle, "Iraqi Christians have Important Mission of 'Forgiveness and Reconciliation'," *Catholic News Agency*, December 28, 2018, https://bit.ly/3QlOQ6x.

of community peace agreements among tribal leaders. A particularly important agreement was one concluded in Tikrit, which opened the way for 390,000 displaced Iraqis to return home and reestablished economic activity in the region. Today the USIP continues to sponsor local reconciliation efforts throughout Iraq, most specifically in Tal Afar and the Nineveh Plain, two areas close to Mosul, a key site of the conflict with ISIS insurgents.[45]

The national government of Iraq, in addition to its proposals for decentralization mentioned above, has focused on preventing violence motivated by revenge. One of its recent initiatives is a High Committee for Coexistence and Communal Peace. In August 2018, in conjunction with the USIP group Sanad for Peacebuilding, this committee helped negotiate a peaceful coexistence settlement in the area of Al-Ayadiyah that "commits tribes to the state's justice and security process instead of tribal justice, which has at times triggered vengeful acts." More than ninety tribes signed the agreement, pledging to work with security forces to track ISIS suspects and to "move beyond parochial identities by giving priority to national identity."[46]

Religious leaders carry particularly strong potential to serve as reconcilers. Parallel to the sense in which the ethic of political reconciliation is inspired and informed by religion yet articulable in secular terms, religious leaders are strongly positioned to promote reconciliation even while they are not its exclusive practitioners. Religious leaders are especially important in Iraq, whose main fault lines fall along religious lines. A prominent example is Shia cleric Ayatollah Ali al-Husseini al-Sistani, who has used the strong respect he holds among Iraqis to call for the cessation of violence, peace between Sunni and Shia communities, the inclusion of Sunnis in the government, and a civil state in which citizenship and office are not based on religious identities.[47] Certain religious leaders from outside Iraq have been important as well. For instance, Anglican Canon Andrew White forged an Iraqi Interreligious Initiative that fosters cooperation between high-level Shiite and Sunni leaders as well as leaders of minority religious groups, including Iraqi Christians.

45 Nancy Lindborg and Sarhang Hamasaaed, "To Stabilize Iraq After ISIS, Help Iraqis Reconcile," United States Institute of Peace, June 27, 2017, https://bit.ly/3voROip.

46 Ibid.; and Sanad for Peacebuilding, "Announcement of Coexistence Pact of Honor for the Tribes in Al-Ayadiyah Sub-District," 2018, https://bit.ly/3bQYdMP.

47 Alaaldin, "Sectarianism, Governance, and Iraq's Future"; and Ali Mamouri, "Shiite Leaders Forbid Insults Against Sunnis," *Al-Monitor: The Pulse of the Middle East,* January 13, 2015, https://bit.ly/3pfiD54.

To suggest that a such a broad array of actors would carry out a wide set of practices of reconciliation is not to assert blithe optimism but rather to point to an approach that can mitigate a most difficult policy dilemma and that embodies a concept of justice and peace that expands upon what the just war tradition envisions. Rooted in religious traditions, purporting to advance moral and political thought, political reconciliation also proffers operational promise. Therein lies its realism.

Bibliography

Abdul-Zahra, Qassim, and Susannah George. "Iraq Holding More than 19,000 Because of ISIS Militant Ties." *Associated Press*, March 21, 2018. https://apnews.com/article/aeece6571de54f5dba3543d91deed381.

Aboueldahab, Noha. "Writing Atrocities: Syrian Civil Society and Transitional Justice." Brookings Doha Center, Analysis Paper Number 21 (May 2018). www.brookings.edu/research/writing-atrocities-syrian-civil-society-and-transitional-justice/.

Abu-Nimer, Mohammed. *Nonviolence and Peace Building in Islam*. Gainesville: University Press of Florida, 2003.

Acharya, Amitav. *The End of American World Order*. Cambridge: Polity Press, 2018.

Acharya, Amitav, and Barry Buzan. *The Making of Global International Relations*. Cambridge: Cambridge University Press, 2019.

Adorno, Theodor W. *Negative Dialectics*. New York: Continuum Press, 1981.

Alaaldin, Ranj. "Sectarianism, Governance, and Iraq's Future." Brookings Institution Doha Center analysis paper no. 24, November 2018. https://brook.gs/3Qpuxp5.

Alexander, Caroline. "Faces of War." *Smithsonian Magazine*. February 2007.

Allison, Graham. *Destined for War: Can America and China Escape the Thucydides Trap?*. New York: Mariner Books, 2017.

American Bar Association. *Criminal Justice Standards for the Prosecution Function*. Washington, DC: American Bar Association, 2017. https://bit.ly/3QBOAQI.

American Bar Association (ABA). 'Rule 3.8: Special Responsibilities of a Prosecutor." *Model Rules of Professional Conduct*. https://bit.ly/3JTSovT.

Anderson, Eric C. *China Restored: The Middle Kingdom Looks to 2020 and Beyond*. Santa Barbara, CA: Praeger, 2010.

Arendt, Hannah. *On Violence*. London: Hardcourt, 1970.

Armitage, David. *Civil Wars: A History in Ideas*. New Haven, CT: Yale University Press, 2017.

Arreguín-Toft, Ivan. *How the Weak Win Wars: A Theory of Asymmetric Conflict*. Cambridge: Cambridge University Press, 2005.

Arthur, Paige. "How 'Transitions' Reshaped Human Rights: A Conceptual History of Transitional Justice." *Human Rights Quarterly*, 31, no. 2 (2009), 321–67.

Austin, John Langshaw. "A Plea for Excuses: The Presidential Address." *Proceedings of the Aristotelian Society*, 57, no. 1 (June 1, 1957), 1–30.

Bacevich, Andrew J. *America's War for the Greater Middle East*. New York: Random House, 2017.

Backer, David. "Victims' Responses to the Truth Commissions: Evidence from South Africa." In Muna Ndulo (ed.), *Security, Reconciliation and Reconstruction: When Wars End*. London: University College of London Press, 2007: 165–96.

Badie, Bertrand. *Humiliation in International Relations: A Pathology of Contemporary International Systems*. Oxford: Hart Publishing, 2017.

Balch-Lindsay, Dylan, and Andrew J. Enterline. "Killing Time: The World Politics of Civil War Duration: 1820–1992." *International Studies Quarterly*, 44, no. 4 (2000), 615–42.

Barnes, Julian E., Thomas Gibbons-Neff, and Eric Schmitt. "Officials Try to Sway Biden Using Intelligence on Potential for Taliban Takeover of Afghanistan." *The New York Times*, April 23, 2021. www.nytimes.com/2021/03/26/us/politics/biden-afghanistan-intelligence.html.

Barnhart, Joslyn. *The Consequences of Humiliation: Anger and Status in World Politics*. New York: Cornell University Press, 2020.

Bass, Gary J. "Jus Post Bellum." *Philosophy & Public Affairs*, 32, no. 4 (October 2004), 384–412.

Bellamy, Alex J. "The Responsibilities of Victory: Jus Post Bellum and the Just War." *Review of International Studies*, 34 (2008), 601–25.

Bellamy, Alex J. *World Peace (And How We Can Achieve It)*. New York: Oxford University Press, 2019.

Benhaida, Sarah. "Head of UN Investigators Calls for an IS 'Nuremberg'." *Yahoo! News*, July 29, 2019. https://yhoo.it/2OqPWCR.

Bica, Camillo C. "A Therapeutic Application of Philosophy." *International Journal of Applied Philosophy*, 13, no. 1 (1999), 81–92.

Blank, Laurie R., and Benjamin R. Farley. "Identifying the Start of Conflict: Conflict Recognition, Operational Realities and Accountability in the Post-9/11 World." *Michigan Journal of International Law*, 36, no. 3 (2015), 496–518.

Borch, Frederic L. *Judge Advocates in Combat: Army Lawyers in Military Operations from Vietnam to Haiti*. Washington, DC: Office of the Judge Advocate General and Center for Military History, 2001.

Boudreau, Tyler. *Packing Inferno: The Unmaking of a Marine*. Minneapolis, MN: Feral House, 2008.

Boudreau, Tyler. "The Morally Injured." *The Massachusetts Review*, 52, no. 3–4 (2011), 746–54.

Brahimi, Lakhdar. "Report of the Panel on United Nations Peace Operations." United Nations, 2000. www.unv.org/publications/report-panel-united-nations-peace-operations-brahimi-report-a55305.

Brittain, Vera. *Testament of Youth*. London: Orion, 2014 [1933].

Brock, Rita, and Gabriella Lettini. *Soul Repair: Recovering from Moral Injury after War*. Boston: Beacon Press, 2013.

Brueggemann, Walter. *The Prophetic Imagination*. Minneapolis, MN: Augsburg Fortress, 2001.

Buford, Warren, and Hugo van der Merwe. "Reparations in Southern Africa." *Cahiers d'etudes africaines*, 44, no. 1/2 (2004), 2–53.

Bybee, Jay S. "Authority of the President Under Domestic and International Law to Use Military Force against Iraq." *Opinions of the Office of Legal Counsel*, 26, (2002), 1–48. https://fas.org/irp/agency/doj/olc/force.pdf.

Callard, Agnes. *Aspiration*. New York: Oxford University Press, 2018.

Camus, Renaud. *You Will Not Replace Us*. Plieux: Chez l'auteur, 2018.

Camus, Renaud. *Le Grand Remplacement*. Plieux: Chez l'auteur, 2019.

Carlson, John D. "Discerning Justice in the Trial and Execution of Saddam Hussein." In David K. Linnan (ed.), *Enemy Combatants, Terrorism, and Armed Conflict Law: A Guide to the Issues*. Westport, CT: Praeger Security International, 2008: 307–26.

Carr, Edward Hallett. *The Twenty Years' Crisis 1919–1939: An Introduction to the Study of International Relations*. London: Palgrave, 1939 [2016].

Carranza, Ruben, Cristián Correa, and Elena Naughton. *More than Words: Apologies as a Form of Reparation*. New York: International Center for Transitional Justice, 2015.

Chairman of the Joint Chiefs of Staff. "Standing Rules of Engagement/Standing Rules for the Use of Force." Instruction 3121.01B. June 13, 2005.

Chivers, C. J. *The Fighters: Americans in Combat in Afghanistan and Iraq*. New York: Simon & Schuster, 2018.

Cicero, Marcus Tullius. *Cicero on the Emotions: Tusculan Disputations 3 and 4*. M. Graver (ed.), Chicago, IL: University of Chicago Press, 2002.

Cicero, Marcus Tullius. *Cicero on the Emotions*. Chicago, IL: University of Chicago Press, 2009.

Cohen, Gerald Allan. *Rescuing Justice and Equality*. Cambridge, MA: Harvard University Press, 2008.

Collier, Paul. *Breaking the Conflict Trap: Civil War and Development Policy*. Washington, DC: The World Bank, 2003.

Congressional Research Service. "American War and Military Operations Casualties: Lists and Statistics." July 29, 2020. https://crsreports.congress.gov/product/pdf/RL/RL32492/33.

Cooley, Alexander, and Daniel Nexon. *Exit from Hegemony: The Unraveling of the American Global Order*. New York: Oxford University Press, 2020.

Council on Foreign Relations. "The Iraq War: 2003–2011." www.cfr.org/timeline/iraq-war.

Council on Foreign Relations. "The U.S. War in Afghanistan: 1999–2021." www.cfr.org/timeline/us-war-afghanistan.

Cox, Walter T., et al., "Report of the Commission on the 50th Anniversary of the Uniform Code of Military Justice." November 4, 2001. https://bit.ly/3SJzZEJ.

Cropper, Maureen L., Sema K. Aydede, and Paul R. Portney. "Rates of Time Preference for Saving Lives." *The American Economic Review*, 82, no. 2 (1992), 469–72.

Cunliffe, Philip. *The New Twenty Years' Crisis: A Critique of International Relations 1999–2019*. Montreal: McGill-Queen's University Press, 2020.

Cunningham, David E. "Veto Players and Civil War Duration." *American Journal of Political Science*, 50, no. 4 (2006), 875–92.

Cushman, Fiery, Kurt Gray, Allison Gaffey, and Wendy Berry Mendes. "Simulating Murder: The Aversion to Harmful Action." *Emotion*, 12, no. 1 (2012), 2–7.

Daniels, Joe Parkin. "Colombia Tribunal Reveals at Least 6,402 People were Killed by Army to Boost Body Count." *The Guardian*, February 19, 2021. www.theguardian.com/global-development/2021/feb/19/colombia-farc-tribunal-false-positives.

D'Arms, Justin, and Daniel Jacobson. "The Moralistic Fallacy: On the 'Appropriateness' of Emotions." *Philosophy and Phenomenological Research*, 61, no. 1(2000), 65–90.

D'Arms, Justin, and Daniel Jacobson. "Sentiment and Value." *Ethics*, 110, no. 4 (2000), 722–48.

Darwall, Stephen. *The Second-Person Standpoint*. Cambridge, MA: Harvard University Press, 2009.

De Brabandere, Eric. "The Responsibility for Post-Conflict Reforms: A Critical Assessment of Jus Post Bellum as a Legal Concept." *Vanderbilt Journal of Transnational Law*, 43, no. 1 (2010), 119.

Deigh, John. *From Psychology to Morality*. New York: Oxford University Press, 2018.

Democracy Now. "Exclusive: Tomas Young Reads in Full His Letter to Bush & Cheney, 'A Message From a Dying Veteran.'" March 21, 2003. https://bit.ly/3vWzDAN.

Department of Defense. *DoD Executive Agent for Non-Lethal Weapons (NLW), and NLW Policy*. Glossary. Directive 3000.03E. Washington, DC: Department of Defense, 2013.

Department of Defense. *DoD Law of War Program*, Directive 2311.01E. Washington, DC: Department of Defense, 2006.

Descartes, Rene. *Meditations on First Philosophy*. Indianapolis, IN: Hackett Publishing, 1993 [1641].

Diamond, Todd. "The Role of Decentralization in Combatting Extremist Influence in Iraq." *The Atlantic Council*, 2017. https://bit.ly/3cPoZFx.

Dimeglio, Richard. "The Evolution of the Just War Tradition: Defining Jus Post Bellum." *Military Law Review*, 186 (Winter ed., 2005), 146–62.

Dixon, Jeffrey. "What Causes Civil Wars? Integrating Quantitative Research Findings." *International Studies Review*, 11, no. 4 (2009), 707–35.

Dodge, Toby. "Intervention and Dreams of Exogenous Statebuilding: The Application of Liberal Peacebuilding in Afghanistan and Iraq." *Review of International Studies*, 39, no. 5 (2013), 1189–212.

Dubik, James M. *Just War Reconsidered: Strategy, Ethics, and Theory*. Lexington: University Press of Kentucky, 2016.

Eberle, Christopher J. *Religious Conviction in Liberal Politics*. New York: Cambridge University Press, 2002.

Ehrhart, W. D. "A Relative Thing." http://vietnamwar.lib.umb.edu/experience/docs/poem_WDEhrhart.html.

Eisikovits, Nir. *Sympathizing with the Enemy: Reconciliation, Transitional Justice, Negotiation*. Boston, MA: Brill, 2009.
Eisikovits, Nir. "ISIS, Humiliation and Political Philosophy." *Journal of Military Ethics*, in press.
Eisikovits, Nir. "Transitional Justice." In Edward N. Zalta (ed.), *The Stanford Encyclopedia of Philosophy*. Fall 2017 ed.
el Bunni, Anwar. "Breaking New Ground: Transitional Justice in Syria." Brookings Institute, November 24, 2020. www.brookings.edu/opinions/breaking-new-ground-transitional-justice-in-syria/.
El-Sadat, Anwar. *In Search of Identity*. New York: Harper Collins, 1978.
Evans, Mark. "Moral Responsibilities and the Conflicting Demands of *Jus Post Bellum*." *Ethics & International Affairs*, 23, no. 2 (2009), 147–64.
Fabre, Cécile. "War Exit." *Ethics*, 125, no. 3 (2015), 631–52. https://doi.org/10.1086/679562.
Farrell, Theo. *Unwinnable: Britain's War in Afghanistan: 2001–2014*. London: Vintage, 2017.
Faulkner, William. *A Fable*. New York: Random House, 2011.
Fawcett, Edmund. *Conservatism: The Fight for a Tradition*. Princeton, NJ: Princeton University Press, 2020.
Fazal, Tanisha M. *State Death: The Politics and Geography of Conquest, Occupation and Annexation*. Princeton, NJ: Princeton University Press, 2007
Fazal, Tanisha M., and Ryan D. Griffiths. "Membership Has Its Privileges: The Changing Benefits of Statehood." *International Studies Review*, 16, no. 1 (2014), 79–106.
Fearon, James. "Why Do Some Civil Wars Last So Much Longer than Others?" *Journal of Peace Research*, 41, no. 3 (2004), 275–301.
Fearon, James and David D. Laitin. "Ethnicity, Insurgency, and Civil War." *American Political Science Review*, 97, no. 1 (2003), 75–90.
Felter, Joseph and Jacob Shapiro. "Limiting Civilian Casualties as Part of a Winning Strategy: The Case of Courageous Restraint." *Daedalus*, 146, no. 1 (2017), 44–58.
Forst, Rainer. *The Right to Justification*. New York: Columbia University Press, 2011.
Frankfurt, Harry. *The Importance of What We Care About*. New York: Cambridge University Press, 1988: 80–94.
Franklin, Benjamin. "On the Criminal Laws and the Practice of Privateering." Letter to Benjamin Vaughan, March 14, 1785. In *The Works of Dr. Benjamin Franklin*. Nurnberg and New York: N. Hickman, 1835.
Freeman, Mark. *Necessary Evils*. New York: Cambridge University Press, 2009.
Frevert, Ute. *The Politics of Humiliation*. New York: Oxford University Press, 2020.
Fukuyama, Francis. *Identity: The Demand for Dignity and the Politics of Resentment*. New York: Farrar, Straus and Giroux, 2018.
Fuller, Lon Luvois. *The Morality of Law*, rev. ed. New Haven, CT: Yale University Press, 1969.
Fussell, Paul. *The Great War and Modern Memory*. New York: Oxford University Press, 1975.

Gabbay, Zvi D. "Justifying Restorative Justice: A Theoretical Justification for the Use of Restorative Justice Practices." *Journal of Dispute Resolution*, 2, (2005), 349–98.

Gerges, Fawaz A. *ISIS: A History*, rev. ed. Princeton, NJ: Princeton University Press, 2017.

Gheciu, Alexandra, and Jennifer Welsh. "The Imperative to Rebuild: Assessing the Normative Case for Postconflict Reconstruction." *Ethics & International Affairs*, 23, no. 2 (2009), 121–46.

Gibbs, Philip. *Now It Can Be Told*. Scotts Valley, CA: CreateSpace Independent Publishing Platform, 2014.

Gilbertson, Ashley. "The Life and Lonely Death of Noah Pierce." *The Utne Reader*, February 13, 2009. www.utne.com/politics/lonely-death-noah-pierce-ptsd-iraq-war.

Goldsmith III, Jack L. "'Protected Person' Status in Occupied Iraq Under the Fourth Geneva Convention." *Opinions of the Office of Legal Counsel*, 28, (2004), 35–61. https://bit.ly/3QmKLzg.

Gordon, Neil. "Misconduct and Punishment." *Center for Public Integrity*, June 26, 2003. https://bit.ly/3dtPody.

Graver, Margaret. *Stoicism and Emotion*. Chicago, IL: University of Chicago Press, 2008.

Gray, Matt, William Nash, and Brett Litz. "When Self-Blame Is Rational and Appropriate: The Limited Utility of Socratic Questioning in the Context of Moral Injury." *Cognitive and Behavioral Practice*, 24, no. 4 (2017), 383–87.

Greenwood, Christopher. "Historical Development and Legal Basis." In Dieter Fleck (ed.), *The Handbook of Humanitarian Law in Armed Conflicts*. New York: Oxford University Press, 1999: 101–50.

de Greiff, Pablo. *The Handbook of Reparations*. New York: Oxford University Press, 2008.

de Greiff, Pablo and Roger Duthie (eds.). *Transitional Justice and Development: Making Connections*. New York: Social Science Research Council, 2009.

Griffin, Brandon J., Everett L. Worthington, Don E. Davis, Joshua N. Hook, and Shira Maguen. "Development of the Self-Forgiveness Dual-Process Scale." *Journal of Counseling Psychology*, 65, no. 6 (November 2018), 715–26.

Griffin, Brandon, Marilyn Cornish, Shira Maguen, Everett L. Worthington, Jr. "Forgiveness as a Mechanism of Repair following Military-Related Moral Injury." In J. Currier, J. Nieuwsma, and K. Drescher (eds.), *Addressing Moral Injury in Clinical Practice*. Washington, DC: APA Publishing, 2019: 71–86.

Griswold, Charles. *Forgiveness: A Philosophical Exploration*. New York: Cambridge University Press, 2007.

Grossman, Dave. *On Killing: The Psychological Cost of Learning to Kill in War and Society*. New York: Back Bay Books, 1995.

Hamilton, Alexander. "Federalist No. 70." In Clinton Rossiter (ed.), *The Federalist Papers* (New York: Penguin, 2003).

Hayner, Priscilla B. *Unspeakable Truths: Transitional Justice and the Challenge of Truth Commissions*. New York: Routledge, 2010

Hegre, Håvard, Mokleiv Nygård, and Ranveig Flaten Ræder. "Evaluating the Scope and Intensity of the Conflict Trap." *Journal of Peace Research*, 54, no. 2 (2017), 243–61.

Heinemann, Larry. *Paco's Story*. New York: Farrar, Straus and Giroux, 1984.

Hellsten, Sirkku. "Transitional Justice and Aid." UNU/WIDER Working Paper 2012/06, UNU/Helsinki. www.wider.unu.edu/sites/default/files/wp2012-006.pdf.

Hendin, Herbert, and A. L. Haas. "Suicide and Guilt as Manifestations of PTSD in Vietnam Combat Veterans." *The American Journal of Psychiatry*, 148, no. 5 (1991), 586–91.

Hieronymi, Pamela. "The Wrong Kind of Reason." *The Journal of Philosophy*, CII (2005), 437–57.

Hieronymi, Pamela. "I'll Bet You Think this Blame Is About You." In Justin Coates and Neal Tognazzini (eds.), *Oxford Studies in Agency and Responsibility Vol. 5: Themes from the Philosophy of Gary Watson*. Oxford: Oxford University Press, 2019: 60–87.

Hill, Thomas E. "Making Exceptions without Abandoning the Principle: Or How a Kantian Might Think about Terrorism." In Raymond Gillespie Frey and Christopher Armstrong (eds.), *Violence, Terrorism, and Justice*. New York: Cambridge University Press, 1991: 196–229.

Hirst, Michael. "Northern Ireland Riots: What Is Behind the Violence in Northern Ireland?" *BBC News,* April 14, 2021. www.bbc.com/news/uk-northern-ireland-56664378.

Hobbes, Thomas. *Leviathan*. Indianapolis, IN: Hackett Publishing, 1994 [1651].

Holmes, Robert L. *On War and Morality*. Princeton, NJ: Princeton University Press, 2014.

Human Rights Watch. "Amicus Brief on Killings of Human Rights Defenders in Colombia." April 20, 2021. www.hrw.org/news/2021/04/20/amicus-brief-killings-human-rights-defenders-colombia.

Huntington, Samuel P. *The Soldier and the State: The Theory and Politics of Civil-Military Relations*. Cambridge, MA: Harvard University Press, 2003 [1957].

Hurka, Thomas. "Proportionality and the Morality of War." *Philosophy and Public Affairs*, 33, no. 1 (2004), 34–66.

Iasiello, Louis V. "*Jus Post Bellum*: The Moral Responsibilities of Victors in War." *Naval War College Review*, LVII (2004), 40–51.

Idaho Supreme Court. "Idaho Criminal Rule 18.1." https://isc.idaho.gov/icr18-1.

"Investigations and Police: Overview of the Fourth Amendment." *Georgetown Law Journal Annual Review of Criminal Procedure*, 45 (2016), 250–69.

Jacoby, Russell. *Bloodlust: On the Roots of Violence from Cain and Abel to the Present*. New York: Free Press, 2011.

Jervis, Robert. "Liberalism, the Blob, and American Foreign Policy: Evidence and Methodology." *Security Studies*, 19, no. 3 (2020), 434–56.

Kamm, F. M. *Ethics for Enemies*. Oxford: Oxford University Press, 2011.

Kane, Tim. *Global US Troop Deployments: 1950–2005*. Washington, DC: Heritage Institute, 2006.

Kant, Immanuel. *Perpetual Peace and Other Essays on Politics, History, and Morals*, translated by Ted Humphrey. Indianapolis, IN: Hackett Publishing, 1983.

Kant, Immanuel. *Ethical Philosophy*. 2nd ed. Indianapolis, IN: Hackett Publishing, 1994.

Kant, Immanuel. *The Science of Right*, translated by W. Hastie. Telangana State, India: A and D Publishing, 2018 [1790].

Kathman, Jacob D. "Civil War Contagion and Neighboring Interventions." *International Studies Quarterly*, 54, no. 4 (2010), 989–1012.

Kellogg, Davida E. "Jus Post Bellum: The Importance of War Crimes Trials." *Parameters*, 32, no. 3 (2002), 87–99.

Khazan, Olga. "Masks: The Face Transplants of World War I." *Atlantic Magazine*. August 2014. www.theatlantic.com/health/archive/2014/08/the-first-face-transplants-were-masks/375527/

Kimmerling, Baruch and Joel Migdal. *The Palestinian People: A History*. Cambridge, MA: Harvard University Press, 2009.

King Jr., Martin Luther. "Beyond Vietnam." speech delivered April 4, 1967. https://bit.ly/3zgpTlY.

King Jr., Martin Luther. *A Testament of Hope: The Essential Writings of and Speeches of Martin Luther King, Jr.* New York: Harper Collins, 1990.

Konchalovsky, Andrei. Interview with *Vesti Nedeli*. January 24, 2009. www.vesti.ru/article/2162310.

Kovic, Ron. *Born on the Fourth of July*. Brooklyn, NY: Akashic Books, 2005.

Krems, J. A., K. E. G. Williams, A. Atipis, and D. T. Kenrick. "Friendship Jealousy: One Tool for Maintaining Friendships in the Face of Third-Party Threats?" *Journal of Personality and Social Psychology*, 120, no. 4 (2021), 977–1012.

Krstic, Branislav, and Fatos Bytyci. "Kosovo Serbs and NATO Troops Clash in Tense North." *Reuters*, June 12, 2012. www.reuters.com/article/us-kosovo-serbs-idUSBRE8500HV20120601.

Lazar, Seth. *Sparing Civilians*. New York: Oxford University Press, 2015.

Lazar, Seth. "Moral Sunk Costs." *The Philosophical Quarterly*, 68, no. 273 (2018), 841–61.

Lazar, Seth. "War." In Edward N. Zalta (ed.), *The Stanford Encyclopedia of Philosophy*. Spring 2020 ed.

Lazare, Aaron. *On Apology*. New York: Oxford University Press, 2004.

Leconte, Cécile. "The Socio-Political Career of the Expression 'The Great Replacement' Among Right-Wing Party Networks in Germany: The Case of the Alternative for Germany (AfD) party." *Politix*, 126, no. 2 (2019), 111–34.

Lederach, John Paul. *Building Peace: Sustainable Reconciliation in Divided Societies*. Washington, DC: United States Institute of Peace, 1997.

Lindborg, Nancy and Sarhang Hamasaaed. "To Stabilize Iraq After ISIS, Help Iraqis Reconcile." United States Institute of Peace, June 27, 2017. https://bit.ly/3voROip.

Lindner, Evelin. *Making Enemies: Humiliation and International Conflict*. Santa Barbara, CA: Praeger, 2006.

Lindner, Evelin. *Honor, Humiliation, and Terror: An Explosive Mix and How to Diffuse It with Dignity*. New York: Dignity Press, 2017.

Litz, Brett T., Nathan Stein, Eileen Delaney, Leslie Lebowitz, William P. Nash, Caroline Silva, and Shira Maguen. "Moral Injury and Moral Repair in War Veterans: A Preliminary Model and Intervention Strategy." *Clinical Psychology Review*, 29, no. 8 (2009), 695–706.

Litz, Brett T., Leslie Lebowitz, Matt. J. Gray, and William Nash. *Adaptive Disclosure: A New Treatment for Military Trauma, Loss, and Moral Injury*. New York and London: Guilford Press, 2016.

Lo, Bobo. *Russia and the New World Disorder*. Washington, DC: Brookings Institute Press, 2015.

Long, A. A., D. N. Sedley, and David M. Sedley. *The Hellenistic Philosophers: Volume 1, Translations of the Principal Sources with Philosophical Commentary*. New York: Cambridge University Press, 1987.

Lu, Catherine. *Justice and Reconciliation in World Politics*. Cambridge: Cambridge University Press, 2017.

Maguen, Shira, et al., "The Impact of Killing in War on Mental Health Symptoms and Related Functioning." *Journal of Traumatic Stress*, 22, no. 5 (2009), 435–43.

Maguen, Shira, et al., "The Impact of Reported Direct and Indirect Killing on Mental Health Symptoms in Iraq War Veterans." *Journal of Traumatic Stress*, 23, no. 1 (2010), 86–90.

Maguen, Shira, et al., "The Impact of Killing on Mental Health Symptoms in Gulf War Veterans." *Psychological Trauma*, 3, no. 1 (2011), 21–26.

Maguen, Shira, et al., "Killing in Combat May Be Independently Associated with Suicidal Ideation." *Depression and Anxiety*, 29, no. 11 (2012), 918–23.

Maguen, Shira, et al., "Killing and Latent Classes of PTSD Symptoms in Iraq and Afghanistan Veterans." *Journal of Affective Disorders*, 145, no. 3 (2013), 344–48.

Maguen, Shira and Brett Litz. "Moral Injury in Veterans of War." *PTSD Research Quarterly, National Center for Posttraumatic Stress Disorder*, 23, no. 1 (2012), 1–3.

Mallinder, Louise. "Amnesties in the Pursuit of Reconciliation, Peacebuilding, and Restorative Justice." In Jennifer Llewellyn and Daniel Philpott (eds.), *Restorative Justice, Reconciliation, and Peacebuilding*. New York: Oxford University Press, 2014: 138–73.

Mamdani, Mahmood. "Amnesty or Impunity? A Preliminary Critique of the Report of the Truth and Reconciliation Commission of South Africa." *Diacritics*, 32, no. 3/4 (2002), 33–59.

Mamouri Ali. "Shiite Leaders Forbid Insults Against Sunnis." *Al-Monitor: The Pulse of the Middle East*. January 13, 2015. https://bit.ly/3pfiD54.

Mansoor, Peter R. *Surge*. New Haven, CT: Yale University Press, 2013.

Marin, Peter. "Living in Moral Pain." In Robert Emmet Meagher and Douglas Pryer (eds.), *War and Moral Injury: A Reader*. Eugene, OR: Cascade Books, 2018 [1981]: 172–84.

Marshall, Christopher D. *Beyond Retribution: A New Testament Vision for Justice, Crime and Punishment*. Grand Rapids, MI: Wm. B. Eerdmans Publishing, 2001.

Matheson, Michael J. "Remarks on the United States Position on the Relation of Customary International Law to the 1977 Protocols Additional to the 1949 Geneva Conventions at the Sixth Annual American Red Cross-Washington College of Law Conference on International Humanitarian Law (January 22, 1987)." *American University Journal of International Law and Policy*, 2, no. 2 (1987), 415–538.

Maurer, Daniel. "Military Mediation as Military Justice?: Conjectures on Repairing Unit Cohesion in the Wake of Relational Misconduct." *Ohio State Journal on Dispute Resolution*, 28, no. 2 (2013), 433–38.

Maurer, Daniel. "The Judge Advocate as Strategist." *Small Wars Journal*, 2017.

McAuley, James. "How Gay Icon Renaud Camus Became the Ideologue of White Supremacy." *The Nation*, June 17, 2019. https://bit.ly/3zQOHB3.

McAuliffe, Padraig. "Transitional Justice and the Rule of Law: The Perfect Couple or Awkward Bedfellows?" *Hague Journal on the Rule of Law*, 2 (2010), 127–54.

McCrae, John. "In Flanders Fields." www.poetryfoundation.org/poems/47380/in-flanders-fields.

McCready, Doug. "Ending the War Right: Jus Post Bellum and the Just War Tradition." *Journal of Military Ethics*, 8, no. 1 (2009), 66–78.

McMahan, Jeff. *Killing in War*. Oxford: Oxford University Press, 2009.

McMahan, Jeff. "Proportionality in the Afghanistan War." *Ethics & International Affairs*, 25, no. 2 (2011), 143–54.

McMahan, Jeff. "Proportionality and Time." *Ethics*, 125, no. 3 (2015), 696–719.

Mearsheimer, John J. *The Great Delusion: Liberal Dreams and International Realities*. New Haven, CT: Yale University Press, 2018.

Mejía, Camilo. *Road from ar Ramadi: The Private Rebellion of Staff Sergeant Mejia: An Iraq War Memoir*. New York: New Press, 2007.

Mejía, Camilo. "Healing Moral Injury: A Lifelong Journey." *Fellowship*, 76 (2011), 25–27.

Menon, Shivshankar. "How China Bucked Western Expectations and What It Means for World Order." Brookings Order from Chaos blog. March 10, 2016. https://brook.gs/3vZc8al.

Miller, Richard W. "The Ethics of America's Afghan War." *Ethics & International Affairs*, 25, no. 2 (2011), 103–31.

Miller, Seamus. "Social Institutions." In Edward N. Zalta (ed.), *The Stanford Encyclopedia of Philosophy*. 2014.

Mishra, Pankaj. *Age of Anger: A History of the Present*. New York: Macmillan, 2017.

Moellendorf, Darrel. "Is the War in Afghanistan Just?" *Imprints a Journal of Analytical Socialism*, 6, no. 2 (2002).

Moellendorf, Darrel. "*Jus Ex Bello*." *Journal of Political Philosophy*, 2, no. 2 (2008), 123–36.

Moellendorf, Darrel. "*Jus Ex Bello* in Afghanistan." *Ethics & International Affairs*, 25, no. 2 (2011), 155–64.

Moellendorf, Darrel. "Two Doctrines of *Jus Ex Bello*." *Ethics*, 125, no. 3 (2015), 653–73.

Moffett, Luke. "Reparations in Transitional Justice: Justice or Political Compromise?" Queen's University Belfast School of Law, Research Paper 2019–07 (2017), https://pureadmin.qub.ac.uk/ws/portalfiles/portal/120688204/Reparations_in_Transitional_Justice.pdf.

Molendijk, Tine. "Toward an Interdisciplinary Conceptualization of Moral Injury: From Unequivocal Guilt and Anger to Moral Conflict and Disorientation." *New Ideas in Psychology*, 51 (December 2018), 1–8.

Molendijk, Tine, Eric-Hans Kramer, and Désirée Verweij. "Moral Aspects of 'Moral injury': Analyzing Conceptualizations on the Role of Morality in Military Trauma." *Journal of Military Ethics*, 17, no. 1 (2018), 36–53.

Müller-Markus, Christina. "One Belt, One Road: The Chinese Dream and Its Impact on Europe." *Notes Internacionals CIDOB*, 148 (2016), 1–6.

Murdoch, Simon and Joe Mulhall. *From Banners to Bullets: The International Identitarian Movement*. London: HOPE not hate Charitable Trust, 2019. https://bit.ly/3QgXK5t.

Murphy, Colleen. *A Moral Theory of Political Reconciliation*. Cambridge: Cambridge University Press, 2010.

Murphy, Colleen. *The Conceptual Foundations of Transitional Justice*. Cambridge: Cambridge University Press, 2017.

Murphy, Colleen. "On Principled Compromise: When Does a Process of Transitional Justice Qualify as Just?" *Proceedings of the Aristotelian Society*, CXX (2020), 47–70.

Murphy, Colleen and Charles Fogelman. "Gender, Structural Inequality, and Just Governance." In Uchendu Chigbu (ed.), *Land Governance and Gender: The Tenure-Gender Nexus in Land Management and Land Policy*. Boston: CABI International, in press.

Murphy, Colleen and Linda Radzik. "Reconciliation." In Edward N. Zalta (ed.), *The Stanford Encyclopedia of Philosophy*. Winter 2020 ed.

Nabulsi, Karma. *Traditions of War: Occupation, Resistance, and the Law*. Oxford: Oxford University Press, 1999.

Nagel, Thomas. *Mortal Questions*. New York: Cambridge University Press, 1979.

Nalepa, Monika. *Skeletons in the Closet: Transitional Justice in Post-Communist Europe*. New York: Cambridge University Press, 2010.

Nash, William P., and Brett T. Litz. "Moral Injury: A Mechanism for War-Related Psychological Trauma in Military Family Members." *Clinical Child and Family Psychology Review*, 16, no. 4 (2013), 365–75.

Nordieimer, Jon. "From Dakto to Detroit: Death of a Troubled Hero." *The New York Times*, May 26, 1971. www.nytimes.com/1971/05/26/archives/from-dakto-to-detroit-death-of-a-troubled-hero-from-dakto-to.html.

Norman, Richard. *Ethics, Killing and War*. Cambridge: Cambridge University Press, 1995.

Norman, Sonya B., and Shira Maguen. "Moral Injury." National Center for PTSD. www.ptsd.va.gov/professional/treat/cooccurring/moral_injury.asp.

North Atlantic Treaty Organization (NATO). *Glossary of Terms and Definitions*. AAP-6. Brussels: NATO Standardization Office, 2019.

Nozick, Robert. *Anarchy, State, and Utopia.* New York: Basic Books, 1974.
O'Brien, Tim. *The Things They Carried.* Boston and New York: Houghton Mifflin, 1980.
O'Driscoll, Cian. *Victory: The Triumph and Tragedy of Just War.* New York: Oxford University Press, 2019.
Orakhelashvili, Alexander. "The Interaction Between Human Rights and Humanitarian Law: Fragmentation, Conflict, Parallelism, or Convergence?" *European Journal of International Law*, 19, no. 1 (2008), 161–82.
Orend, Brian. "*Jus Post Bellum.*" *Journal of Social Philosophy*, 31, no. 1 (2000), 117–37.
Orend, Brian. "Justice after War." *Ethics & International Affairs*, 16, no. 1 (2002), 43–56.
Orend, Brian. *War and International Justice: A Kantian Perspective.* Waterloo, Ontario: Wilfrid Laurier University Press, 2006.
Orend, Brian. "*Jus Post Bellum*: Perspectives of a Just-War Theorist." *Leiden Journal of International Law*, 20, no. 3 (2007), 579–91.
Orend, Brian. *The Morality of War*, 2nd ed. Peterborough, Ontario: Broadview Press, 2013.
Orwell, George. *The Lion and the Unicorn: Socialism and the English Genius.* London: Penguin, 2018.
Paine, Thomas. "Agrarian Justice." In Moncure Daniel Conway (ed.), *The Writings of Thomas Paine, Volume III. 1791–1804.* New York and London: G. P. Putnam's Sons, 1895: p. 191.
Parfit, Derek. *Reasons and Persons.* Oxford: Oxford University Press, 1984.
Parfit, Derek. *On What Matters.* Oxford: OxfoErd University Press, 2011.
Parks, W. Hays "The United States Military and the Law of War: Inculcating an Ethos." *Social Research*, 69, no.4 (2002), 981–1015.
Paust, Jordan J. "Dr. Francis Lieber and the Lieber Code." *Proceedings of the Annual Meeting (American Society of International Law)*, 95 (2001), 112–15.
Peace Vet. "Personal Legacy of the War." December 17, 2011. www.svaphilosopher.com/Peersonal-Legacy-of-War.html.
Pham, Phuong N., Nehal Bhuta, Patrick Vinck, and Nida Alahmad. *Iraqi Voices: Attitudes Toward Transitional Justice and Social Reconstruction.* Berkeley: The International Center for Transitional Justice and the Human Rights Center at the University of California, 2004.
Philpott, Daniel. "Religious Freedom and Peacebuilding: May I Introduce You Two?" *The Review of Faith & International Affairs*, 11, no. 1 (2013), 31–37.
Philpott, Daniel. *Just and Unjust Peace: An Ethic of Political Reconciliation.* New York: Oxford University Press, 2015.
Philpott, Daniel. *Religious Freedom in Islam: The Fate of a Universal Human Right in the Muslim World Today.* New York: Oxford University Press, 2019.
Pickering, Timothy. *An Easy Plan of Discipline for a Militia.* Salem, MA: Samuel and Ebenezer Hall, 1775. https://bit.ly/3zJ4VoA.
Plato. *Symposium.* Translated by Alexander Nehamas and Paul Woodruff. Indianapolis, IN: Hackett Publishing, 1989.
Porter, Patrick. *The False Promise of Liberal Order: Nostalgia, Delusion and the Rise of Trump.* Cambridge: John Wiley & Sons, 2020.

Pozner, Vladimir. Interview with *Argumenty i Fakty*. February 4, 2015. https://aif.ru/politics/russia/1438503.

Pufendorf, Samuel. *On the Duty of Man and Citizen According to Natural Law*. James Tully (ed.). Cambridge: Cambridge University Press, 1991.

Putin, Vladimir. "Russia at the Turn of the Millennium." *Nezavisimaia Gazeta*. Translated by Alan Zebek. December 30, 1999.

Putin, Vladimir. "Open Letter to Voters." Kremlin.ru. February 25, 2000. http://en.kremlin.ru/events/president/transcripts/24144.

Rabinowicz, Wlodek, and Toni Rønnow-Rasmussen. "The Strike of the Demon: On Fitting Pro-attitudes and Value." *Ethics*, 114, no. 3 (2004), 391–423.

Ramsey, F. P. "A Mathematical Theory of Saving." *The Economic Journal*, 38, no. 4 (December 1928), 543–59.

Rawls, John. *The Law of Peoples: With the Idea of Public Reason Revisited*. Cambridge, MA: Harvard University Press, 1999.

Rawls, John. *A Theory of Justice*, rev. ed. New York: Oxford University Press, 1999.

Recchia, Stefano. "Just and Unjust Postwar Reconstruction: How Much External Interference Can Be Justified?" *Ethics & International Affairs*, 23, no. 2 (2009), 165–87.

Refugee Law Project. "Forgiveness: Unveiling an Asset for Peacebuilding." 2015. https://bit.ly/3JgN3wV.

Regan, Patrick M. "Third Party Interventions and the Duration of Interstate Conflicts." *Journal of Conflict Resolution*, 46, no. 1 (2002), 55–73.

Remarque, Erich Maria. *All Quiet on the Western Front*. New York: Ballantine Books, 1987.

Remarque, Erich Maria. *All Quiet on the Western Front*. New York: Everyman's Library, 2018.

Revkin, Mara Redlich. *The Limits of Punishment: Transitional Justice and Violent Extremism*. Tokyo: United Nations University, 2018. https://bit.ly/2zI6nQC.

Robeyns, Ingrid, and Morten Fibieger Byskov. "The Capability Approach." In Edward N. Zalta (ed.), *The Stanford Encyclopedia of Philosophy* (Winter 2020 ed.). https://plato.stanford.edu/entries/capability-approach/

Rodin, David. "The War Trap: Dilemmas of *Jus Terminatio*." *Ethics*, 125, no. 3 (2015), 674–95.

Rothberg, Michael. *The Implicated Subject: Beyond Victims and Perpetrators*. Stanford, CA: Stanford University Press, 2019.

Rothstein, Richard. *The Color of Law: A Forgotten History of How Our Government Segregated America*. New York: Liveright Publishing, 2017.

Rousselle, Christine. "Iraqi Christians Have Important Mission of 'Forgiveness and Reconciliation'." *Catholic News Agency*, December 28, 2018. https://bit.ly/3QlOQ6x.

Sachs, Susan. "Iraqis Seek Justice, Or Vengeance, for Victims of the Killing Fields." *The New York Times*, November 4, 2003.

Sadat, Anwar. *In Search of Identity: An Autobiography*. New York: Harper Collins, 1978.

Sanad for Peacebuilding. "Announcement of Coexistence Pact of Honor for the Tribes in Al-Ayadiyah Sub-District." 2018. https://bit.ly/3bQYdMP.

Sandoval, Clara and Miriam Puttick. "Reparations for the Victims of Conflict in Iraq: Lessons Learned from Comparative Practice." Ceasefire Centre for Civilian Rights and Minority Rights Group International. 2017. https://bit.ly/3zlzRCA.

Scanlon, Thomas. *What We Owe to Each Other*. Cambridge, MA: Harvard University Press, 1988.

Schaap, Andrew. *Political Reconciliation*. London: Routledge, 2004.

Schmitt, Michael N. "Investigating Violations of International Law in Armed Conflict." *Harvard National Security Journal*, 2, no. 1 (2011), 31–84.

Schindler, Dietrich. "Human Rights and Humanitarian Law: Interrelationship of the Laws." *American University Law Review*, 31, no.4 (1982), 935–41.

Schorr, Yonit, et al., "Sources of Moral Injury Among War Veterans: A Qualitative Evaluation." *Journal of Clinical Psychology*, 74, no. 12 (2018), 1–16.

Schuck, Michael J. "When the Shooting Stops: Missing Elements in the Just War Theory." *Christian Century*, 111, no. 30(1994), 982–84.

Scranton, Roy and Matt Gallagher (eds.), *Fire and Forget: Short Stories from the Long War*. Boston: Da Capo Press, 2013.

Seamone, Evan. "Reclaiming the Rehabilitative Ethic in Military Justice: The Suspended Punitive Discharge as a Method to Treat Military Offenders with PTSD and TBI and Reduce Recidivism." *Military Law Review*, 208, no. 1 (2011), 1–212.

Seneca. *On Mercy*. vol. 1. J. W. Basore (ed.), Cambridge, MA: Harvard University Press, 1985.

Seneca, Lucius Annaeus. *Six Tragedies*. Emily Wilson (ed.), Oxford: Oxford University Press, 2010.

Sharp, Dustin. "What Would Satisfy Us? Taking Stock of Critical Approaches to Transitional Justice." *International Journal of Transitional Justice*, 13, no. 3 (2019), 570–89.

Shatan, Chaim F. "Post-Vietnam Syndrome." *The New York Times*. May 6, 1972. www.nytimes.com/1972/05/06/archives/postvietnam-syndrome.html

Shay, Jonathan. *Odysseus in America: Combat Trauma and the Trials of Homecoming*. New York: Scribner, 2002.

Shay, Jonathan. *Achilles in Vietnam: Combat Trauma and the Undoing of Character*. New York: Simon & Schuster, 2010.

Shay, Jonathan. "Moral Injury." *Psychoanalytic Psychology*, 31, no. 2 (2014), 182–91.

Shay, Jonathan. "Moral Leadership Prevents Moral Injury." In Robert Emmet Meagher and Douglas Pryer (eds.). *War and Moral Injury: A Reader* (Eugene, OR: Cascade Books, 2018 [2010]), 301–6.

Shelby, Tommie. "Ideology, Racism, and Critical Social Theory." *The Philosophical Forum*, 34, no. 2 (2003), 153–88.

Sherman, Nancy. *Stoic Warriors: The Ancient Philosophy Behind the Military Mind*. New York: Oxford University Press, 2005.

Sherman, Nancy. *The Untold War: Inside the Hearts, Minds, and Souls of Our Soldiers*. New York: W. W. Norton & Company, 2010.

Sherman, Nancy. *The Untold War: Inside the Hearts, Minds, and Souls of Our Soldiers*. New York: Oxford University Press, 2010.

Sherman, Nancy. *Afterwar: Healing the Moral Wounds of Our Soldiers*. New York: Oxford University Press, 2015.

Sherman, Nancy. *Stoic Wisdom: Ancient Lessons for Modern Resilience*. New York: Oxford University Press, 2021.

Shoemaker, David. *Responsibility from the Margins*. New York: Oxford University Press, 2015.

Sheehan, Cindy. *Peace Mom: A Mother's Journey Through Heartache to Activism*. New York: Atria Books, 2010.

Sierra Leone Truth and Reconciliation Commission. "Witness to Truth: Final Report of the TRC." 2004. www.sierraleonetrc.org/index.php/view-the-final-report/download-table-of-contents/volume-one.

Silkin, Jon (ed.), *Out of Battle: The Poetry of the Great War*. London: Palgrave Macmillan, 1998.

Sissons, Miranda, and Abdulrazzaq Al-Saiedi. "Iraq's De-Baathification Still Haunts the Country." *Al Jazeera*, March 12, 2013. www.aljazeera.com/opinions/2013/3/12/iraqs-de-baathification-still-haunts-the-country.

Sissons, Miranda, and Abdulrazzaq Al-Saiedi. "A Bitter Legacy: Lessons of De-Baathification in Iraq." International Center for Transitional Justice Report. March 4, 2013. www.ictj.org/sites/default/files/ICTJ-Report-Iraq-De-Baathification-2013-ENG.pdf.

Slote, Michael. "Morality and Self-Other Asymmetry." *The Journal of Philosophy*, LXXXI, no. 4 (1984), 179–92.

Slote, Michael. *From Morality to Virtue*. New York: Oxford University Press, 1995.

Snyder, Timothy. "The American Abyss." *The New York Times*, January 9, 2021. www.nytimes.com/2021/01/09/magazine/trump-coup.html.

Solis, Gary D. *The Law of Armed Conflict*. New York: Cambridge University Press, 2017

South African Truth and Reconciliation Commission. "Final Report." 1998. www.justice.gov.za/trc/.

Stahn, Carsten. "*Jus ad Bellum*', '*jus in bello*' … '*jus post bellum*'? – Rethinking the Conception of the Law of Armed Force." *European Journal of International Law*, 17, no. 5 (2006), 921–43.

Stanton, Jessica A. *Violence and Restraint in Civil War: Civilian Targeting in the Shadow of International Law*. New York: Cambridge University Press, 2016.

Steinhauer, Jennifer. "Suicides Among Post-9/11 Veterans Are Four Times as High as Combat Deaths, a New Study Finds." *The New York Times*, June 22, 2021. www.nytimes.com/2021/06/22/us/911-suicide-rate-veterans.html.

Sterling, Joe. "In Iraq, Truth Commission Idea Gains Traction." *CNN*, March 25, 2009. https://cnn.it/3JTB4WH.

Strachan, Hew. *The Direction of War: Contemporary Strategy in Historical Perspective*. New York: Cambridge University Press, 2013.

Strawson, P. F. "Freedom and Resentment." *Proceedings of the British Academy*, 48 (1962), 187–211.

Supreme Court of Ohio. *Ohio Rules of Professional Conduct (as amended)*. Columbus: Supreme Court of Ohio, 2020. https://bit.ly/3bTYdLN.

Supreme Court of Pennsylvania. *Rules of Professional Conduct*. Harrisburg: Supreme Court of Pennsylvania. 2019. https://bit.ly/3QkUi9L.

Supreme Court of the State of Illinois. *Illinois Rules of Professional Conduct*. 2010. https://bit.ly/3JUmWfu.

Tadros, Victor. *To Do, To Die, To Reason Why: Individual Ethics in War*. Oxford: Oxford University Press, 2020.

Tessman, Lisa. *Moral Failure: On the Impossible Demands of Morality*. New York: Oxford University Press, 2015.

Tessman, Lisa. *When Doing the Right Thing Is Impossible*. New York: Oxford University Press, 2017.

Tessman, Lisa. "The Virtues of Reactive Attitudes." *The Journal of Value Inquiry*, 55, no. 3 (2021), 437–56.

The Iraq Study Group. *The Iraq Study Group Report*. New York: Random House, 2006.

The Irish Times. "Iraqi Prime Minister Unveils Reconciliation Plan." June 26, 2006. https://bit.ly/3Pn8iP5.

Thomason, Krista K. "Guilt and Child Soldiers." *Ethical Theory and Moral Practice*, 19, no. 1 (2015), 115–27.

Thucydides. *The History of the Peloponnesian War*. Translated by Rex Warner. London: Penguin Classics, 1972.

Thucydides. *The Landmark Thucydides*. Robert Strassler (ed.), New York: Touchstone, 1998.

Tooby, John and Leda Cosmides. "Friendship and the Banker's Paradox: Other Pathways to the Evolution of Adaptations for Altruism." In W. G. Runciman, John Maynard Smith, and R. I. M. Dunbar (eds.), *Evolution of social behaviour patterns in primates and man*. Oxford and New York: Oxford University Press, 1996: 119–43.

Tripathi, Deepak. *Imperial Designs: War, Humiliation & the Making of History*. Washington, DC: Potomac Books, 2013.

Trumbo, Dalton. *Johnny Got His Gun*. New York: Citadel, 2007.

United Nations. "Guidance Note of the Secretary General: United Nations Approach to Transitional Justice." March 2010. https://digitallibrary.un.org/record/682111?ln=en

United Nations Secretary General. "Supplement to an Agenda for Peace." 1995. https://digitallibrary.un.org/record/168325?ln=en

United Nations Security Council. "Resolution 1244." U.N. Doc. S/RES/1244. June 10, 1999. https://undocs.org/S/RES/1244(1999).

United Nations Special Rapporteur on Truth, Justice, Reparations, and Guarantees of Non-Recurrence. "Memorialisation Processes: Report." July 9, 2020. https://documents-dds-ny.un.org/doc/UNDOC/GEN/G20/175/70/PDF/G2017570.pdf?OpenElement.

United States Army Field Manual. "Legal Support to the Operational Army." FM 1–04. 2013. https://bit.ly/3JevWfn.

United States Census Bureau. "Population, Population Change, and Estimated Components of Population Change: April 1, 2010 to July 1, 2020." Census.gov. https://bit.ly/3QvoLyY.

US Army. *Operational Law Handbook*, 18th ed. Charlottesville, VA: The Judge Advocate General's Corps Legal Center and School, 2018.

US Department of Army. *Rules of Professional Conduct for Lawyers*. Washington, DC: United States Department of the Army, 2018.

US Department of Army. *Field Manual 6–27, The Commander's Handbook on the Law of Land Warfare*. Washington, DC: Headquarters, Department of the Army, 2019.

US Department of Defense. *Legal Support to Military Operations, GL-3*. Joint Publication (JP). Washington, DC: United States Department of Defense, 2011: 1–04.

US Department of Defense. *Department of Defense Dictionary of Military and Associated Terms (as amended)*. Joint Publication (JP) Washington, DC: Department of Defense, 2016: 1–02.

US Department of Defense. *Law of War Manual*. Washington, DC: Office of General Counsel, United States Department of Defense, 2016.

US Department of Defense. *United States Manual for Courts-Martial*. Washington, DC: United States Department of Defense, 2019.

US Department of Justice. *Justice Manual*. Title 9, § 9–27.000, "Principles of Federal Prosecution." www.justice.gov/jm/jm-9-27000-principles-federal-prosecution.

Van Antwerpen, Jonathan. "Reconciliation as Heterodoxy." In Jennifer Llewellyn and Daniel Philpott (eds.), *Restorative Justice, Reconciliation, and Peacebuilding*. New York: Oxford University Press, 2014: 77–117.

Von Einseidel, Sebastian. "Civil War Trends and the Changing Nature of Armed Conflict." *U.N. University Centre for Policy Research*, Occasional Paper 10, 2017.

Verdeja, Ernesto. *Unchopping a Tree: Reconciliation in the Aftermath of Political Violence*. Philadelphia, PA: Temple University Press, 2009.

Vice News. "Bulldozing the Border Between Iraq and Syria: The Islamic State." Online video clip. August 14, 2014. www.youtube.com/watch?v=TxX_THjtXOw.

Walker, Shaun. "The Humiliation that Pushed Putin to Try and Recapture Russian Glory." *History.com*, February 23, 2018. www.history.com/news/vladimir-putin-russia-power.

Walker, Shaun. *The Long Hangover: Putin's New Russia and the Ghosts of the Past*. New York: Oxford University Press, 2018.

Walt, Stephen M. *The Hell of Good Intentions: America's Foreign Policy Elite and the Decline of US Primacy*. New York: Farrar, Straus and Giroux, 2018.

Walzer, Michael. *Just and Unjust Wars: A Moral Argument with Historical Illustrations*. New York: Basic Books, 1977.

Walzer, Michael. "The Triumph of Just War Theory (and the Dangers of Success)." *Social Research*, 69, no. 4 (2002), 925–44.

Walzer, Michael. *Arguing About War*. New Haven, CT: Yale University Press, 2004.

Walzer, Michael. "Just and Unjust Occupations." *Dissent*, 51, no. 1 (2004), 61–63.

Walzer, Michael. "The Aftermath of War: Reflections on Jus Post Bellum." In Eric Patterson (ed.), *Ethics Beyond War's End*. Washington, DC: Georgetown University Press, 2012: 35–46.

Washington, George. "Address to the New York Provincial Congress." June 26, 1775 https://founders.archives.gov/documents/Washington/03-01-02-0019.

Watson, Gary. *Agency and Answerability: Selected Essays*. Oxford: Clarendon Press, 2004.

Watson, Gary. "Peter Strawson on Responsibility and Sociality." In David Shoemaker and Neal Tognazzini (eds.), *Oxford Studies in Agency and Responsibility, Vol II: 'Freedom and Resentment' at 50*. Oxford: Oxford University Press, 2014: 15–32.

Weil, Simone. *Selected Essays*. Eugene, OR: Wipf and Stock, 1962.

Weisbach, David A., and Cass R. Sunstein. "Climate Change and Discounting the Future: A Guide for the Perplexed." *Yale Law & Policy Review*, 27, no. 2 (2009), 433–57.

Wertheim, Stephen. *Tomorrow, the World*. Cambridge, MA: Belknap Press of Harvard University, 2020.

West Virginia Judiciary. *Rules of Professional Conduct*. Rule 3.8(d). Charleston, SC: West Virginia Judiciary. 2015, https://bit.ly/3w1zw72.

Wildman, Sarah. "'You Will Not Replace Us': A French Philosopher Explains the Charlottesville Rant." *Vox*, August 15, 2017. https://bit.ly/3QDaDq6.

Williams, Bernard. *Ethics and the Limits of Philosophy*. Cambridge, MA: Harvard University Press, 1985.

Williams Jr., Robert E. and Dan Caldwell. "*Jus Post Bellum*: Just War Theory and the Principles of Just Peace." *International Studies Perspectives*, 7, no. 4 (2006), 309–20.

Wilson, Emily. *The Greatest Empire: A Life of Seneca*. New York: Oxford University Press, 2014.

Witt, John Fabian. *Lincoln's Code: The Laws of War in American History*. New York: Free Press, 2012.

Wolf, Susan. "The Moral of Moral Luck." *Philosophic Exchange*, 31 (2000), 1–16.

Wong, David B. *Natural Moralities: A Defense of Pluralistic Relativism*. New York: Oxford University Press, 2006.

Wood, David. *What Have We Done: The Moral Injury of Our Longest Wars*. New York: Little, Brown, 2016.

Young, Iris Marion. *Justice and the Politics of Difference*, rev. ed. Princeton, NJ: Princeton University Press, 2011.

Yu, Hong. "Motivation behind China's 'One Belt, One Road' Initiatives and Establishment of the Asian Infrastructure Investment Bank." *Journal of Contemporary China*, 26, no. 105 (2017), 353–68.

Zemach, Ariel. "The Unpleasant Responsibilities of International Human Rights Law." *Denver Journal of International Law & Policy*, 38, no. 3 (2010), 101–36.

Zheng Wang. *Never Forget National Humiliation: Historical Memory in Chinese Politics and Foreign Relations*. New York: Columbia University Press, 2014.

Zicheng, Ye. *Inside China's Grand Strategy*. Lexington: University Press of Kentucky, 2011.

Zucchino, David. "The War in Afghanistan: How It Started and How It Is Ending." *New York Times*, July 31, 2021. www.nytimes.com/article/afghanistan-war-us.html.

Index

For EU product safety concerns, contact us at Calle de José Abascal, 56–1°,
28003 Madrid, Spain or eugpsr@cambridge.org.

www.ingramcontent.com/pod-product-compliance
Lightning Source LLC
LaVergne TN
LVHW011812200926
848484LV00005B/1168

* 9 7 8 1 1 0 8 8 3 4 2 8 5 *